D0291259

ECOCRITIQUE

ECOCRITIQUE

CONTESTING THE POLITICS OF NATURE, ECONOMY, AND CULTURE

Timothy W. Luke

University of Minnesota Press
Minneapolis
London

Copyright 1997 by the Regents of the University of Minnesota

All rights reserved. No part of this publication may be reproduced, stored in a retrieval system, or transmitted, in any form or by any means, electronic, mechanical, photocopying, recording, or otherwise, without the prior written permission of the publisher.

Published by the University of Minnesota Press
111 Third Avenue South, Suite 290, Minneapolis, MN 55401-2520
Printed in the United States of America on acid-free paper

Library of Congress Cataloging-in-Publication Data

Luke, Timothy W.
     Ecocritique : contesting the politics of nature, economy, and
culture / Timothy W. Luke.
        p.   cm.
     Includes index.
     ISBN 0-8166-2846-7 (alk. paper). — ISBN 0-8166-2847-5 (pbk. : alk.
paper)
     1. Environmentalism.   2. Green movement.   I. Title.
GE195.L85   1997
363.7—dc21   97-12049

The University of Minnesota is an
equal-opportunity educator and employer.

To Nikki

Contents

Acknowledgments

For more than thirty years, the environmental ills of advanced industrial societies, like those of Europe, Japan, or North America, have been poked and prodded by a diverse band of ecological critics. Posing as eco-philosophers, ecowarriors, ecopoliticians, ecoengineers, or ecoplanners, these "ecocritics" all have advanced their own unique "ecocritiques" as the correctives needed to check the spreading plague of environmental decline. Such ecocritiques constitute an important discursive tradition of critical ecological theory and environmental analysis, which can, in turn, be reevaluated to appraise how effectively green resistance movements have contested the politics of nature, economy, and culture in America since the 1960s.

This book carefully reconsiders the workings of some well-known environmental action groups, as well as the writings of several significant ecological critics, to draw a more complete picture of their political and economic agendas. As a critical rereading of ecocritiques, I implicitly elaborate my own style of ecocritique by engaging other ecological critics and environmental resistance movements in a political debate over theories and practices. Each chapter articulates how I have come to understand several important environmental movements, groups, and thinkers as they have operated in the American context since the mid-1980s. Some sections have appeared elsewhere in different forms. Chapter 1 is a revised version of an article in *Telos* 76 (summer 1988), and chapter 2 elaborates on a paper from *Current Perspectives in Social Theory* 14 (1994). Chapter 3 appeared in *Capitalism Nature Socialism* 6, no. 2 (June 1995), and another version of chapter 4 was first published in *Capitalism Nature*

Socialism 5, no. 2 (June 1994). Chapter 5 draws from a longer and different article in the *Ecologist* 25, no. 4 (July–August 1995). Chapter 6 revises a chapter from *In the Nature of Things: Language, Politics, and the Environment*, ed. Jane Bennett and William Chaloupka (Minneapolis: University of Minnesota Press, 1993). An earlier version of chapter 7 first appeared in *Marcuse: From the New Left to the Next Left*, ed. John Bokina and Timothy J. Lukes (Lawrence: University Press of Kansas, 1994). Portions of chapter 8 were published in *Telos* 100 (fall 1994), and chapter 9 combines a few passages from articles in *Social Science Journal* 24 (June 1987) and *Telos* 88 (summer 1991). Each chapter has been considerably reworked and revised for this book.

During the years in which I have developed these critiques, a number of people have given me a great deal of assistance on some part of the manuscript or all of it. In particular, Ben Agger, Jane Bennett, John Bokina, Bill Chaloupka, Gary Downey, John Ely, Ellsworth R. Fuhrman, Suzi Gablik, Barbara Lawrence, Timothy J. Lukes, James O'Connor, Paul Taggart, Gearóid ÓTuathail, Doug Taylor, and Edward Weisband all were quite helpful. My colleague Stephen K. White once again provided invaluable comments, as did two very significant critics and editors at large: Paul Piccone and Florindo Volpacchio. At the University of Minnesota Press, Lisa Freeman and Carrie Mullen both provided tremendous support for this project.

The material tasks of generating the text itself were done with extraordinary skill and care by Kim Hedge, Terry Kingrea, and Maxine Riley in the Department of Political Science at Virginia Polytechnic Institute and State University. I am most thankful to them for dealing with my continual changes and for their good work.

Introduction: Contesting the Politics of Nature, Economy, and Culture

The Environment. Silent Spring. Embattled Nature. Our Ecology. Population Explosion. Damaged Ecosystems. Mother Earth. Terms such as these have turned into anchor points for social critics, moral philosophers, and policy experts over the past three decades. Because nothing in Nature simply is given within society, such terms must be assigned significance by every social group that mobilizes them as meaningful constructs. As a result, a never-ending flow of moral arguments, cultural quarrels, and policy squabbles constantly collide with various constructions of nature, economy, and culture deployed in the political discourses of any state and society.

Many styles of ecologically grounded criticism circulate in present-day American mass culture, partisan debate, consumer society, academic discourse, and electoral politics as episodes of ecocritique, contesting our politics of nature, economy, and culture in the contemporary global system of capitalist production and consumption.[1] As these debates unfold, visions of what is the good or bad life, where right conduct or wrong action for individuals repose, how progress should or should not be realized, and why solidarity or estrangement might grip communities increasingly find many of their most compelling articulations as "ecocritiques."[2] Ecocritique has become a common genre of analysis mobilized for and against various projects of power and economy in the organization of our everyday existence.[3]

As it has evolved since the early 1960s, the full range of environmental analysis, nature philosophy, or ecological criticism now is quite vast. Few studies of its many complexities do justice either to all of its various discursive manifestations or to each of these traditions' most representative voices.[4] This book is not an exhaustive survey of every ideological variation and political permutation in recent ecological criticism. Instead, it assembles one series of closely focused treatments of particular thinkers, movements, or groups that have been somewhat ignored or neglected in other considerations of ecological criticism. All of the ecological critics addressed here provide important examples of contemporary ecocritique, but their peculiar ideological orientations or political commitments frequently have sidelined them in previous discussions, strangely enough, for being either too radical, marginal, and extreme or too mainstream, established, and ordinary to merit critical consideration. By turning their positions over in my rereadings, I begin constructing my own critique by challenging them on how they would have us reimagine our politics of nature, economy, and culture. Still, my rereadings only provide one more perspective, which is neither absolutely right nor ultimately fail-safe, on these diverse schools of ecological theorizing. As critiques of ecological critics, my readings should elicit other critical readings of my insights and arguments. This outcome is to be expected as an integral part of contesting the politics of nature, economy, and culture in the existing world system of unsustainable development.

Going beyond more conventional "green" politics, I reappraise how power and economy, society and culture, community and technology operate for and against what are now widely regarded as the embattled ecosystems of Nature.[5] Beginning with deep ecology and concluding with social ecology, this book first looks at a number of ecological movements and environmental groups before reconsidering a few individual thinkers with green philosophical leanings. My ecocritique then centers on how contemporary ecocritics read "ecology," taking their discourse seriously on its own terms as the study of the totality of all interrelations between a human society with everything in its environment. Instead of accepting those relationships as they are now conceptualized through metaphors of biosystemic equilibrium, my ecocritiques rethink ecology by exploring the totality of all human/machine, human/animal, human/plant interactivities as power/knowledge relations, and argue that there are many other alternative forms for creating our built environments, high technologies, and economic communities. Building the concrete

expressions of such alternatives would reconstitute the our nature/
economy/culture equations, materially and symbolically, without per-
petuating much of the ecological destruction that mars their operations
today.[6] By reworking the practices of our economies and polities, new
industrial metabolisms, fresh process aesthetics, alternative technology
regimes might find a place in material existence beyond the simplicities
of either radical anthropocentrism or fundamentalist biocentrism in
survivable communitarian ecologies within which people dominate neither other human beings nor their fellow nonhuman beings.

This orientation is apparent from the outset as chapter 1 attempts to
take the teachings of deep ecology seriously as a political philosophy.
Rather than exiling it to the reservation of "radical ecology," like many
recent examinations of its doctrines, this reading of deep ecology sees it
as a significant philosophical system and social force that needs to be
appraised out in the open on its own merits.[7] The efforts of Arne
Naess, Bill Devall, George Sessions, and others to articulate an ethics for
deep ecology are among the more challenging attempts to rethink the
politics of nature, economy, and culture in the current world system of
global capitalism. Nonetheless, their project is full of flaws that have
hobbled their social movement's political appeal as well as curtailed their
philosophical system's environmental insights. Deep ecology's unusual
fetishization of wilderness, as well as its anachronistic fascination with
mythologized preagricultural peoples, often couple, in particular, its philosophical precepts with a crippling set of counterproductive individual
actions and problematic social values.

This inclination in deep ecology is discussed further in chapter 2,
which provisionally reviews the organization and operation of Earth First!
The ecocritiques of Earth First! are—perhaps somewhat perversely—an
interesting example of a modern philosophy of praxis concocted by an
unlikely group of organic intellectuals operating underground or out
in the field as an environmental resistance movement. Although the core
membership of Earth First! has been quite small in numbers and continually shaken by internal divisions over its methods, goals, values, and
leaders, it continues to be an active force in many localities, fighting for
the preservation of wilderness and the abolition of industrialism. Much
of its organizational mythology, on the one hand, is an almost situation-

ist enactment of an ecodystopian novel by Edward Abbey, *The Monkey Wrench Gang*, while most of its institutionalized practices, on the other hand, depend on exploiting high technology's internal contradictions against itself.[8] However, the movement limits its chance for greater strategic success inasmuch as it celebrates revitalizing not merely ecosystems and human cultures that prevailed during the preindustrial era, but rather those that existed in preagricultural times.[9]

Another sort of preservationistic revivalism is reconsidered in chapter 3, in which the work of The Nature Conservancy is studied closely as another style of practically engaged ecocritique. Although it prefers to operate well outside of the mass media limelight, The Nature Conservancy has worked for more than four decades to build its own vast assemblage of nature preserves all over North and South America by soliciting and investing funds from private donors. Although the achievements of The Nature Conservancy are now quite tangible inasmuch as "Nature" is conserved as real estate, the Conservancy's procedures for mounting its environmental resistance to global capitalism by organizing a war of position out of leveraging clever real estate deals are proving, unfortunately, to be not all that resistant. Instead of ardently opposing the destruction of Nature in general, The Nature Conservancy seems content with conserving small pieces of undeveloped land to preserve tiny bits and pieces of habitat as precious containers of biodiversity. As a result, building a Nature Conservancy by using capitalist strategies is more akin to maintaining a "nature cemetery" than truly preserving Nature from capitalism.

Chapter 4 extends the doubts raised about a mainstream, almost establishmentarian, group like The Nature Conservancy to the work of Lester Brown and his Worldwatch Institute. The radical ecology of Earth First! pales to insignificance when contrasted to the rootedness of the Worldwatch Institute's environmental surveillance systems as a serious environmentalist agency. This chapter, however, worries about seeking this kind of "credibility." Indeed, it seems to be a clear sign of a dangerous new tendency shared by many contemporary ecocritiques; that is, the ecological criticism generated by groups like the Worldwatch Institute is now aimed not so much at advancing the protection of Nature and liberation of society from global commerce as toward enhancing the managerial strategies exercised by governmental and nongovernmental regulatory organizations over transnational business as private firms continue advancing their exploitation of Nature. Consequently, as chapter 4 suggests, the Worldwatch Institute mostly operates as another integral part of the emergent alliances of big businesses, nongovernmental organizations, and global think tanks that have been collaborating in the invention of new discourses for a global "governmentality," articulated now through the disciplinary categories of "sustainable development."[10]

The high technology apotheosis of sustainability ideologies undoubtedly has been best represented by the New Age technoscience experiments at Gaia simulation in Biosphere 2. As one almost totalitarian attempt to reinvent and rationalize the ecology of planet Earth as an environmental simulation, the Biosphere 2 project takes the notions of sustainability to new heights as a radical anthropocentrism. Chapter 5 reexamines the initial dealings of the Biosphere 2 project in order to tease out some of the antiecological assumptions of its Nature modeling, which appear to have been oriented more toward inventing techniques for environmentally imperializing extraterrestrial regions with new high-tech terraforming schemes rather than preserving some inner native balance in the still wild zones of the earth's many ecologies. This project's imagination of nature, economy, and culture as an amalgam of biospheric science and tourist infotainment also provides a fascinating look at how fairly radical forms of ecocritique can be reconstructed by corporate capital as leisure industry sites or technoscience theme parks.

In contrast, a low technology incarnation of sustainability ideologies, as chapter 6 indicates, can be discovered in the practices of green consumerism. Some sustainable development discourses attempt to outline systems for reproducing the material culture of advanced industrial society by closing some of the waste and entropy loops in the abstract machines of social production and consumption. These reengineering efforts can take many forms, but closing irrational gaps between the practices of mass consumption in a throwaway society by training "green consumers" increasingly is favored by environmental activists around the world. Just as the throwaway society was invented four decades ago by retraining consumers to accept the products of everyday life in "no deposit, no return" packaging, so too can a new type of sustainable society—the authors of environmental shopping handbooks reason—arise from their intense instruction in green consumerism. This chapter closely reads a number of popular consumer manuals for green consumers in North America, first, to call the project of green consumerism into question, and, second, to challenge the pretensions of revolutionary transformation routinely written into these discourses of environmental protection. At best, the ecocritiques of green consumerism provide another level of weak, constrained artificial negativity to check some of the irrational tendencies of corporate capitalism, but they hardly constitute a radical base for truly transforming consumer societies as they work today.[11]

Chapter 7 shifts the analysis focus away from broader movements and specific groups to begin addressing individual theorists. Here the ecological dimensions of Herbert Marcuse's critical project are reevaluated. Marcuse's analysis of contemporary advanced industrial society in *One Dimensional Man* in many ways is still a vital benchmark for ecological criticism.[12] His systematic critique of how science and technology constitute an abstract machine of subordination, forcing everyday life to suit the dominating imperatives of capital and instrumental reason, while this order also restrains an emancipatory potential for reconciling humanity and its environment in nondestructive liberating modes of production and consumption, in many ways has still not been answered by his detractors. Marcuse often is not seen as an ecological thinker, but much of his work is a theoretical study in the process aesthetics or cultural technics of high technology's embeddedness in Nature.[13] While some might dismiss him as anthropocentric, he should, in fact, be seen as exploring alternative ways of conflating/infusing/compounding the interactivities of humans and machines, humans and animals, or humans and plants in the restructuring of Nature by society. His vision of a "new sensibility" shows a more humane technics-regime embracing Nature in peaceful, cooperative, nurturing terms rather than the aggressive, competitive, dominating categories currently imposed by the prevailing practices of corporate capitalism. Marcuse might be regarded as one of the first contemporary ecological critics to begin rethinking human/natural identities in the "pacification of Nature." In turn, this project would reposition the subjectivities of both an emancipated humanity and other animate beings as they share the same ecosystems linked by new aesthetics, new technics, new ethics beyond the profit-driven dictates of contemporary capital's instrumental rationality.

Chapter 8, in turn, reviews the ecological criticism of Paolo Soleri, whose visionary fusion of architectural planning with environmental alarmism in the new discipline of "arcology" presents a promising, albeit underdeveloped, avenue for organizing ecological politics. By centering the design, construction, and managing of cities at the core of the nature, economy, culture equation, Soleri's ecocritique questions "who" defines "for whom" the material interrelationships of humanity and Nature, people and machines, society and economy under current conditions of production. His own experimental community of Arcosanti in central Arizona mostly dodges the tough cultural, economic, legal, political, and social questions implicitly posed by this call to rebuild cities from the ground up. Still, Soleri's concept of arcology can help us see how existing environmental irrationalities might be thoroughly transformed by communities of ecologically minded citizens.

The decisive importance of restructuring community and revitalizing citizenship in an ecological society is highlighted in chapter 9, which reconsiders the critical analysis of contemporary society advanced by Murray Bookchin's social ecology. Despite the flaws in Bookchin's own ecocritique, he clearly has shown how society cannot be seen as separate and apart from Nature.[14] Moreover, the daunting prospect of reinventing current technologies as ecotechnologies, or technics and tools suited to recentering human/machine, human/animal, human/plant interactions in low-impact, sustainable, nonalienating, beautiful, high-efficiency, localistic forms of productive cooperation, is identified by Bookchin as a key strategy for reinventing human communities as confederalist municipalities without the domination of people by people that has led to the domination of Nature by humanity.

The conclusion weaves together insights from the preceding chapters into a preliminary outline for imagining that social world we need to create beyond the environmental destruction that each of these ecocritiques deplores. To take advantage of the energies mobilized by an Earth First! or Nature Conservancy, and to avoid the pitfalls associated with sustainable development or green consumerism, this alternative, as

Marcuse, Soleri, and Bookchin emphasize, stresses the importance of re-building local communities within a restructured global economy. Lo-calistic communities, which could embed their economies and societies in Soleri's arcological structures to reorder the built environment and re-balance human communities with their natural environments, also have a good shot at developing some of Marcuse's "new sensibility" of them-selves, society, and Nature as they bring various new ethics of peace and beauty to Bookchin's ecotechnological and ecopopulistic means for pacifying their existence.

The real threat to ecological sustainability in our global ecology perhaps is not growth per se, as the Worldwatch Institute, Earth First! or even Soleri, might argue. Resource shortages, when and where they do emerge in contemporary societies, are not the mystical product of "growth," but rather arise from serious irrationalities, as Marcuse, Soleri, and Bookchin contend, in globally articulated modes of industrial, agri-cultural, and informational production that stress profit maximization, or short-run calculations of costs and benefits. Immediately enjoyable commercial utility derived from the management of the earth's ecologies trades mainly private economic benefit off against mostly public ecologi-cal cost.[15] A particular form of transnational corporate commerce sys-tematically structures most of these outcomes. And, its workings consti-tute the most concrete threat to ecological sustainability and natural resources, not some disembodied abstraction known as "growth."

As green consumerism suggests, the key contradictions between Na-ture and the economy are not those conflicts conventionally discovered between "growth" and "ecology" or "jobs" and "the environment." The corporate-backed "wise use" movement often touts these trade-offs as the key contradictions in contemporary political life, but its analysis simplistically assumes that more "growth" and "jobs" always means less "ecology" and "environment."[16] Each of these factors is a function of many complex sets of social decisions, technological choices, political options being exercised in one way rather than another. Pollution oc-curs, resources are wasted, energy is misdirected because of the means that capital, expertise, and government settle on to create, allocate, and expropriate the wealth produced by collective effort and individual ini-tiative out of Nature's many materials. The Nature Conservancy and the

Worldwatch Institute are, in one sense, correct: good jobs, quality growth, unsullied ecologies, and a healthy environment all could be realized simultaneously as new collective goods without returning to Pleistocene-era social institutions and technologies. Population growth can cause environmental degradation, but it also has been a precondition for many social innovations to the extent that "the environment" itself for human beings is simply an abstract artificial assembly in which vast constellations of complex arcological systems extract matter and energy from closely watched, intensely worked, constantly managed, and extensively exploited space.

As chapters 6 and 7 suggest, the environment in twentieth-century society constitutes spaces and provides sites for the assembly and operation of numerous megamachines.[17] These purposive systems for extracting energy out of matter, imposing order on entropy, or realizing structure from information are interlocking more and more humans/animals/plants/minerals/soils in a transnational array of "natural habitats" that support millions of human and nonhuman lives in a massive, diverse, multimorphic constellation of ecological transactions. Capitalist criteria of gains and losses, costs and benefits, growth and decay now are at the heart of much of this ecological deterioration, but these existing habitat systems could just as well conform to less destructive ends of operation, first, without approaching today's unsustainable levels of environmental damage, and, second, still supporting large numbers of human and nonhuman beings.

My ecocritiques rethink how complex interpenetrating ensembles of Nature and society, technology and environment, ecology and economy sustain the earth's myriad forms of human and nonhuman being. "Ecology today," as Worster observes, "is not a single approach to nature; it embraces many approaches." It is important to recognize that, "like the whole of science," ecology continues to be "a house with many doors, some leading to one view of nature, some to another."[18] Ecology drives many different systems of criticism, all intent on generating new social understandings of contemporary human communities' interdependent interrelations with the biosphere. Once we can look beyond the rigid divisions of nature and society or humanity and ecology still used in so many currents of contemporary ecological criticism, we might be able to ask what we want as "the ecological life" from cities and wilderness without compromising the survival or existence of all other nonhuman beings that would make the conditions of this new ecological human being possible.[19]

Learning from the ecocritiques we engage with here, this ecological life may combine elements from both Earth First! and Biosphere 2, but it will neither go back to Pleistocene-era styles of hunting and gathering nor depart on Spaceship Earth missions inside a biospheric cage. Pushing Soleri's insights to their most general level, all of my ecocritiques are attempts to rethink the arcological practices of existing capitalist economies. Indeed, in one sense, this much plainly is shared by all of the many different ecological critics discussed here. Following unquestioningly any of the authors reviewed in these ecocritiques strictly on their own terms, however, soon trips over theoretical tangles twisted into their incomplete appraisal of many present-day political and economic practices. Instead, another modernity must, and can be, found amid the one that now is destructively denaturing Nature, which would be capable of fusing social ecological urbanity and deep ecological humility in social technologies for truly survivable development of human and nonhuman beings.[20]

1

Deep Ecology as Political Philosophy

This chapter reconsiders "deep ecology" as a political philosophy. Deep ecology emerged in the 1970s as a critical reaction to the reform environmentalism of the 1960s, which developed, in turn, as a response to the unfettered exploitation of Nature during the global economic boom of the 1950s and 1960s. Yet, in seeking to improve on reform environmentalism, deep ecologists—such as Arne Naess, Bill Devall, and George Sessions—have taken conceptual positions in their philosophy of nature that are quite problematic. This chapter outlines their basic philosophical stance and then elaborates a critique of some of their more contradictory claims.[1]

The influence of deep ecology extends beyond a new philosophy of nature. Deep ecological principles now motivate many local, bioregional, national, and even transnational political action groups. Numerous "place defense" organizations across America, such as local oppositions to new hydroelectric dams, power lines, highways, timbering programs, dredging and draining schemes, or power plants, tap into deep ecological ideas. Ecological resistance to new nuclear weaponry systems, nuclear power stations, military bases, or communications networks on a local, regional, or national level draws inspiration from deep ecology. Many radical activists, such as the "monkey wrenchers" of the American Southwest and Pacific Northwest, have deep ecological leanings.[2] Similarly, political activist groups, including Earth First!, Greenpeace, Friends of the Earth, the Sea Shepherds, and the various Green political parties in North America, Australia, Western Europe, and Japan, share close affinities with deep ecology.[3] Deep ecology, then, has inspired many new social movements' defense of the quality of their everyday life from the state and transnational commerce during the past decade and a half.[4] As such, it deserves careful consideration.

One can sympathize with much in deep ecology. Most important, it might constitute *in nuce* a new other-regarding code of justice for men and women to deal with each other in society and as society with Nature. By seeing Nature as a significant form of otherness with properties of sentient subjectivity, deep ecology proposes new norms of human responsibility to change the human exploitation of Nature into coparticipation with Nature. However, its practitioners often weaken their appeal by adopting positions that are ineffectively argued or politically naive. Such criticism challenges this significant philosophical project to improve its analysis rather than simply dismissing it *tout court* as morally bankrupt.

State, Society, and the Environment

Modern industrial production, as Schumpeter claims, presumes "creative destruction" in its workings. Vast waste has been a primary product of both modern capitalism and socialism. In the United States, the wasteful destruction of industrial capitalism has sparked at least three major social conservation movements during the twentieth century. The first two took hold within the federal bureaucracy during the Progressive Era and the New Deal period under Theodore and Franklin D. Roosevelt.[5] Both of these movements, however, aimed only at conservation. Their agenda merely limited the most destructive abuses of land, timber, water, and mineral resources in the private sector by charging federal agencies with their scientific management. Economic growth and development were not proscribed; they simply were made somewhat more rational in cost-benefit terms through bureaucratic central monitoring.

During the 1940s and 1950s, environmental conditions changed radically for the worse during the "long prosperity" following Hiroshima. Barry Commoner observes,

> Many pollutants were totally absent before World War II, having made their environmental debut in the war years: smog (first noticed in Los Angeles in 1943), man-made radioactive elements (first produced in

the wartime atomic bomb project), DDT (widely used for the first time in 1944), detergents (which began to displace soap in 1946), synthetic plastics (which became a contributor to the rubbish problem only after the war).[6]

In the United States, the additive impact of these problems sparked critical attention, and then government action in the 1960s on John F. Kennedy's "New Frontier" in a third national conservation movement.

As Sessions claims, the urgent crises of the 1960s split critical thinking about the environment as well as political strategies for its protection.

> Something happened, however, in the mid-1960s which one author describes as a movement "from conservation to ecology." There was a rapidly emerging awareness that increased population, pollution, resource depletion, nuclear radiation, pesticide and chemical poisoning, the deterioration of the cities, the disappearance of wildlife and wilderness, decreases in the "quality of life," and continued economic growth and development under the rhetoric of "progress" had some underlying biological interrelationship. There was a growing suspicion of the ability of technologists to manage natural systems successfully.[7]

A state-sanctioned reform environmentalism was launched during the Kennedy administration by Secretary of the Interior Stewart Udall, while more radical ecological groups reacted both against its moderation and the growing severity of the environmental crisis.[8]

Reform environmentalism and radical ecology movements both focus on the unintended social costs of economic growth, complexity, scale, and productivity. Yet, reform environmentalists treat them as minor problems that can be managed from the public and not-for-profit sector with technocratically planned changes in government regulation or market-driven incentives in the private sector.[9] Most radical ecologists, on the other hand, see the modern service state's approach to these "unintended costs" as *forgotten costs* that business, society, and government have always known about but purposely suppressed. Such costs never can be fully eradicated, because an industrial economy presumes their imposition as externalities. Regulating them only postpones the final reckoning by shifting the costs elsewhere. Radical changes in thinking are needed to attack the more basic problems—untrammeled economic growth, instrumental rationality, and the reification of Nature—implicit in capitalist industrialism. Hence, deep ecologists turn to repressed, ignored, or forgotten visions of ecological living, which persist beneath, behind, or beyond the existing structures of industrial society.

Deep Ecology: Origins and Outlines

Deep ecologists argue that their principles are nothing new. They see themselves borrowing the "ancient truths" of preindustrial, nonurban, and precapitalist societies. Because such societies have been either destroyed or ignored within the current world system, however, deep ecologists identify these basic beliefs as "the minority tradition." Devall and Sessions assert that "the minority tradition focuses on personal growth within a small community and selects a path to cultivating ecological consciousness while protecting the ecological integrity of the place."[10] As examples of such thinking, Sessions tags a panoply of influences from around the world: Christian Franciscanism, Heideggerian philosophy, Aldo Leopold's ecosystem ethics, Taoism, Buddhism, hunter-gatherer tribal religions, Western process metaphysics (Heracleitus, Whitehead, and Spinoza), American Indian culture, European romanticism (Goethe, Rousseau, Blake, Wordsworth, Coleridge, Shelley), American transcendentalism (Emerson, Thoreau, Whitman, Muir), Beat philosophy (Allen Ginsberg and Gary Snyder), 1960s counterculture (Alan Watts, Charles Reich, Theodore Roszak), social ecology (Murray Bookchin, Karl Hess, Duane Elgin), and ecoresistance (John Rodman and Edward Abbey).[11] In one way or another, all of these cultural pieces are placed into the deep ecology conceptual mosaic.

Arne Naess first articulated a sharp distinction during 1972 between "shallow environmentalism" and "deep ecology," which he published in a 1973 article, "The Shallow and the Deep, Long-Range Ecology Movements: A Summary."[12] Naess claims that the conservation movements of the Progressive and New Deal eras, environmental pressure groups, the Club of Rome "limits to growth" school, and animal rights activists all fail to challenge the existing institutionalized worldview of advanced industrial society. Such "shallow" ecologies instead adhere wrongly to an "anthropocentric" view of Nature. This view separates humanity from Nature and deadens it. By seeing Nature as inanimate matter, humans gain the power to dominate it. Working with such values, shallow environmentalism is merely a kind of "enlightened despotism." It only eases, and does not end, the ravages of human domination over Nature by eliminating its worst forms of waste, or regulating its inefficiencies.

Naess's notion of a "deep ecology" accepts shallow environmentalism's intentions, but it pushes beyond this limited approach into a totalistic critique of modern industrialism. He stresses a postanthropocentric "biospherical egalitarianism" to create "an awareness of the equal right (of all things) *to live and blossom.*"[13] By calling for the return to Nature, Naess claims a normative role for deep ecologists in working out an ecosophical approach to Nature. Thus: "By an *ecosophy* I mean a philosophy of ecological harmony or equilibrium. A philosophy is a kind of *sophia* wisdom, is openly normative, it contains both norms, rules, announcements and hypotheses concerning the state of affairs in our universe. Wisdom is policy wisdom, prescription, not only scientific description and prediction."[14] One more systematic version of Naess's deep ecology is Bill Devall and George Sessions's 1985 book *Deep Ecology*, which elaborates the central concepts of deep ecology in much more detail. Following Naess, they see today's basic conflicts as those of consciousness: "Thus deep ecology goes beyond the so-called factual level to the level of self and earth wisdom. Deep ecology goes beyond a limited piecemeal shallow approach to environmental problems and attempts to articulate a comprehensive religious and philosophical worldview. The foundations of deep ecology are the basic intuitions and experiencing of ourselves and Nature which comprise ecological consciousness."[15]

The central problem of the dominant worldview for Devall and Sessions is the human desire to dominate Nature:

> Ecological consciousness and deep ecology are in sharp contrast with the dominant worldview of technocratic-industrial societies which regards humans as isolated and fundamentally separate from the rest of Nature, as superior to, and in charge of, the rest of creation. But the view of humans as separate and superior to the rest of Nature is only part of larger cultural patterns. For thousands of years, Western culture has become increasingly obsessed with the idea of dominance: with dominance of humans over nonhuman Nature, masculine over the feminine, wealthy and powerful over the poor, with the dominance

of West over non-Western cultures. Deep ecological consciousness allows us to see through these erroneous and dangerous illusions.[16]

To acquire this nondominating ecological consciousness, Devall and Sessions adopt Naess's two ultimate norms: self-realization and biocentric equality.

First, self-realization is framed by Gary Snyder's vision of "real work," becoming a whole person rather than an isolated ego struggling to accumulate material possessions.[17] It is a new ethic of "being" or "doing" rather than a credo of "experiencing" or "having." Self-realization is defined as spiritual growth, or the unfolding of inner essence, which "begins when we cease to understand or see ourselves as isolated and narrow competing egos and begin to identify with other humans from our family and friends to, eventually, our species. But the deep ecology sense of self requires a further maturity and growth, an identification which goes beyond humanity to include the nonhuman world."[18]

Second, the norm of biocentrism maintains that "all things have an equal right to live and blossom and to reach their own individual forms of unfolding and self-realization within the larger self-realization." This principle does not preclude mutual predation; instead, it stresses the larger concern of living "with minimal rather than maximal impact on other species and the earth in general."[19] Devall and Sessions argue: "Biocentric equality is intimately related to the all-inclusive Self-realization in the sense that if we harm the rest of Nature then we are harming ourselves. There are no boundaries and everything is interrelated."[20] This awareness, Devall and Sessions maintain, would move people to change their behavior in accord with "voluntary simplicity," or living life as "simple in means, rich in ends."

After defining ecosophy during the 1970s by its opposition to both advanced industrialism and shallow environmentalism, Naess and Sessions developed eight essential principles for deep ecology in 1984: (1) the well-being of human and nonhuman life on earth has intrinsic values, separate from human uses or purposes; (2) the diverse richness of all life forms contributes to realizing these intrinsic values; (3) humans have no right to reduce this richness and diversity of life except to satisfy vital needs; (4) the flourishing of human life and culture is compatible

with a substantial decrease in human populations; indeed, the flourishing of nonhuman life requires such a decrease; (5) human interference
with the nonhuman world is excessive and worsening; (6) policies must
be changed to transform economic, ideological, and technological structures into a situation much different from the present; (7) human satisfaction must shift to appreciating the quality of life (dwelling in situations
of inherent value) rather than adhering to higher material standards of
living; and (8) those who subscribe to these points have an obligation,
directly or indirectly, to try to implement the necessary changes.[21] Although these principles articulate the basic values of deep ecology, they
also contain several flaws that limit their practice.

Conceptual Contradictions in Deep Ecology

Where have the deep ecologists dug their intellectual foundations? In
citing "the minority tradition" as their inspiration, deep ecologists
have combed through the cultural traditions of precapitalist, nonurban,
preindustrial primal peoples, seeking "a basis for philosophy, religion,
cosmology, and conservation practices that can be applied to our own
society."[22] By taking this line, and drawing in notions from previous
Western intellectual advocates of such thinking, deep ecology stands
firmly on one essential point, namely, as a systematic negation of the
"Enlightenment schema."

As Horkheimer and Adorno assert, "the program of the Enlightenment was the disenchantment of the world; the dissolution of myths
and the substitution of knowledge for fancy."[23] The essence of this new
knowledge is instrumental reason qua technology:

> It does not work by concepts and images, by the fortunate insight, but
> refers to method, the exploitation of others' work and capital. . . .
> What men want to learn from Nature is how to use it in order wholly
> to dominate it and other men. That is the only aim. . . . for the En-
> lightenment, whatever does not conform to the rule of computation
> and utility is suspect. . . . Enlightenment is totalitarian.[24]

Horkheimer and Adorno detail how the Enlightenment systematically
extirpated animism and mythological thinking to disenchant Nature. If
any mode of thought failed to attain closure with rational assumptions

and observational judgments, then enlightened consciousness reduced it to mere fictions.

> Enlightenment has always taken the basic principle of myth to be an-
> thropomorphism, the projection onto Nature of the subjective. . . . It
> makes the dissimilar comparable by reducing it to abstract quantities.
> To the Enlightenment, that which does not reduce to numbers, and
> ultimately to the one, becomes illusion; modern positivism writes it
> off as literature. Unity is the slogan from Parmenides to Russell. The
> destruction of gods and qualities alike is insisted upon. . . . myth turns
> into enlightenment, and nature into mere objectivity. Men pay for the
> increases of their power with alienation from that over which they ex-
> ercise their power. Enlightenment behaves toward things as a dictator
> toward men.[25]

Deep ecologists want to overturn this dictatorship of enlightenment, returning human consciousness back to a reenchanted world, an animate
resubjectified Nature, and more mythic modes of knowing to overcome
man's alienation from and domination of Nature. Devall and Sessions
deny enlightened instrumental reasoning any jurisdiction over these
basic precepts: "they are arrived at by the deep questioning process and
reveal the importance of moving to the philosophical and religious level
of wisdom. They cannot be validated, of course, by the methodology of
modern science based on its usual mechanistic assumptions and its very
narrow definition of data."[26] Deep ecology, as Michael Tobias suggests,
concerns "personal moods, values, aesthetic and philosophical convic-
tions which serve no necessarily utilitarian, nor rational end. By definition their sole justification rests upon the goodness, balance, truth and
beauty of the natural world, and of a human being's biological and
psychological need to be fully integrated into it."[27] Given this disposition, deep ecology also draws from primal traditions. "The natural and
supernatural worlds are inseparable; each is intrinsically a part of the
other. Humans and natural entities are in constant spiritual interchange
and reciprocity," because Devall and Sessions interpret the available an-
thropological evidence as indicating that "the primal mind holds the totality of human-centered artifacts, such as language, social organization,
norms, shared meanings, and magic, within the first world of Nature.

For the primal mind there is no sharp break between humans and the rest of Nature."[28]

Deep ecological self-realization, then, is the antithesis of corporate consumerism in which isolated egos strive for more pleasures through their purchasing behaviors. It instead projects this notion of a "primal mind" as a new form of autonomous subjectivity to attain a fresh fusion of human self with the "organic wholeness" of Nature's Self.[29] Both people and Nature are defined as enchanted fields of conscious being, interlocked by natural necessity and human mythic understanding. This process of the "full unfolding of the self can also be summarized by the phrase 'no one is saved until we are all saved,' where the phrase 'one' includes not only me, an individual human, but all humans, whales, grizzly bears, whole rain forest ecosystems, mountains and rivers, the tiniest microbes in the soil, and so on."[30]

In liberating Nature, everything in the biosphere would be treated as an animate subject with inherent rights for self-realization. Thus, "the intuition of biocentric equality is that all things in the biosphere have an equal right to live and blossom and to reach their individual forms of unfolding and self-realization within the larger self-realization. The basic intuition is that all organisms and entities in the ecosphere, as parts of an interrelated whole, are equal in intrinsic worth."[31] People are Nature, and Nature is, at least in part, people. To harm Nature, then, is to commit slow suicide or engage in self-mutilation. When humans technologically intervene in Nature, it means that we assault, abuse, or murder other significant selves from river systems to animals to rain forests. Still, natural subjects are empowered under this rule in Naess's system to use each other as food, shelter, or security, because "mutual predation is a biological fact of life."[32] However, given humanity's tremendous destructive powers in Enlightenment-based technology, people must treat Nature and other natural subjects as ends, not means, living "with minimum rather than maximum impact on other species and on the earth in general." As modes of mythic construction for a reenchanted world, these animistic apperceptions of subjectivity in Nature, Devall and Sessions conclude, "cannot be grasped intellectually but are ultimately experiential."[33]

These arguments do have positive aspects. In particular, they accord human respect to "natural otherness."[34] Deep ecology wants to treat other forms of natural life and nonlife with an ethic of normative responsibility, which might offer a foundation for new systems of justice. Deep ecology also presents substantive rules for enacting this nature consciousness as a mode of human good conscience. Nonetheless, beyond these promising ideas, deep ecology's challenge to the Enlightenment schema is flawed in several respects.

The Myth of Humanity's Fall

The quest for self-realization and biocentric equality in Naess's new ecosophy implicitly assumes its own myth of man's fall. Once upon a time or elsewhere in the world, it is claimed, humanity lived in a state of innocence or grace. But, now or here, due to human technological domination over Nature, humanity mainly lives in a state of corruption or alienation. However, redemption is possible, in accord with the examples set by primal societies, by attaining correct moral consciousness through individual acts of will made imperative by the destruction of Nature. Consequently, the idea of "primal peoples," which in itself is as questionable a rhetorical category as "world communism" or "the Third World," serves as a reified symbol of virtuous praxis.

Deep ecologists toss the many primitive cultures of the world onto one big pile, and then privilege their values and practices—as a register for "the primal mind" as a new subjectivity—unquestioningly in their ecosophy. To stave off ecological crisis in postindustrial society, they recommend that individuals appropriate the norms of preliterate/preindustrial peoples into their everyday activity through acts of goodwill. Supposedly, "primal peoples are characterized by individuation, personalism, nominalism, and existentialism," whereas the pathology of advanced industrialism "consists in our dedication to abstractions, to our collectivism, pseudo-individualism, and lack of institutional means for the expression and transcendence of human ambivalence."[35] Such generalizations are quite suspect, especially without any supporting evidence. Do all primal cultures share these nature-regarding norms as ethical universals? Could primal peoples develop antiecological values? Have primal peoples created ecological crises or abused the environment with their technologies? Does this moral alternative really exist or is it a psychosocial projection discovered in postindustrial social scientists' dissatisfaction with advanced industrialism as they systematize their one-sided understanding of primal people?

Devall and Sessions stress the wisdom and values of American Indians, but this approval reduces a wide range of different cultures with varying values and different practices to a privileged, reified symbol that virtually denies contradiction. Tough questions must be asked. If American Indians are the right model, are we to emulate tribes like the Aztecs and Incas or the Mohaves and Cocopahs? Did any Indian societies systematically practice slavery or war making on their neighbors? Did some groups, like the Hohokam in the pre-Columbian American Southwest, create an ecological disaster by destroying their lands through over-irrigation, much like corporate agriculturalists today? Were all Indian societies equally peaceful, nature-regarding, and respectful of the individual?[36] When deep ecologists claim that primal peoples unfailingly used Nature so that a "richness of ends was achieved with material technology that was elegant, sophisticated, appropriate, and controlled within the context of a traditional society,"[37] red flags must be raised. Such values may have been true for some, but not all, small nomadic bands of hunter-gatherers or slash-and-burn agriculturalists. Can the same rules, however, be followed by ecologizing postindustrial peoples reinhabiting nature? Devall and Sessions deny that they are resurrecting the myth of "noble savages." Still, their search for "inspiration from primal traditions" verges on such a covering myth.

In addition to reducing a broad range of primal cultures into one complex univocal tradition, deep ecologists construct Nature as an active subject that can teach people, if they cultivate their intuition or introspective consciousness, a special redemptive "Earth Wisdom."[38] As Devall and Sessions maintain, "we may not need something new, but need to reawaken something very old, to reawaken our understanding of Earth Wisdom. In the broadest sense, we need to accept the invitation to the dance—the dance of unity of humans, plants, animals, the earth. We need to cultivate an ecological consciousness."[39] Nature is seen as speaking, knowing, having needs, suffering, sharing selfhood, expressing, and growing. Primal traditions are vital because they have remained open to Nature's subjectivity, following its wisdom and sharing in its being. Therefore, deep ecology stresses its antimodern disposition by calling for a reinhabitation of varying bioregions in the future along the primitive lines of primal societies. The myth of humanity's fall from primitive grace arguably is quite false, but this is what justifies deep ecology's antimodern, future primitive vision of social change.

The Dialectic of Reenchantment

In reenchanting the world, deep ecology's privileging of primal traditions ignores the extent to which primal society's myth, magic, and ritual are partially the functional equivalents (and perhaps the conceptual antecedents) of Enlightenment science and technology. Ritual and myth may be the "nondominating science" that Naess, Devall, and Sessions support; yet, as Horkheimer and Adorno argue, these modes of knowing anticipate scientific domination. In mythic reasoning,

> everything unknown and alien is primary and undifferentiated: that which transcends the confines of experience; whatever in things is more than their previously known reality. . . . The dualization of Nature as appearance and sequence, effort and power, which first makes possible both myth and science, originates in human fear, the expression of which becomes explanation. . . . The separation of the animate and the inanimate in the occupation of certain places by demons and deities, first arises from this pre-animism, which contains the first lines of the separation of the subject and object.[40]

Primal myths and rituals also can work as an operationalist mode of thinking, mediating primal people's efforts to control Nature.[41]

"Myth intended report, naming, the narration of the beginning," Horkheimer and Adorno admit, "but also presentation, confirmation, explanation. . . . every ritual includes the idea of activity as a determined process which magic can nevertheless influence. . . . the myths, as the tragedians came upon them, are already characterized by the discipline and power that Bacon celebrated as the 'right mark.'"[42] Magic is based on the specific, not the universal, concrete representation, not substitutional abstractions, and impersonations of demonic or divine spirits, not methodical operational manipulations of matter. It can contain its own dialectic of enlightenment. Mythological ritual sets off the process of enlightenment until enlightenment itself becomes an animistic magic. "Just as myths already realize enlightenment, so enlightenment with every step becomes more deeply engulfed in mythology."[43]

Devall and Sessions's endorsement of the "new physics" of holistic interrelation reflects this confusion.[44] Enlightenment science can not be disinvented or destroyed. It is embedded in all of our acts and artifacts. Sensuous, participative, metaphysical views of reality—when fused into technical potentials of modern science for destructive misapplication—could promote a more domineering rather than less destructive science. There are no guarantees. As the discussion of Biosphere 2 in chapter 5 indicates, New Age thinkers already treat "ecological reasoning" as a model of some automatic circuits for cybernetic management intrinsic to Nature (the "Spaceship Earth" metaphor) that scientists can modify to realize greater micro-macroefficiencies in global production, like weather modification, ocean farming, or genetic engineering.[45] Following this logic, deep ecology can slip too easily into "Whole Earth Catalogism," fusing the quest for more woodstoves, tepees, and windmills with lobbying for interplanetary colonization, space stations, and hypercomputerization.

Deep ecologists forget how much of modern science was itself wrapped up in enchanted, mystical visions of Nature.[46] Newtonian mechanics were created not only to make possible new technical engineering feats, but also to understand the sensuous music of celestial spheres. The unintended consequences of reenchantment and myth may express an operational wish for new instrumental control as much as primal coparticipation in the natural and supernatural worlds. In modern times, this confusing conflation of enchanted mythic ritual and instrumentally rational science perhaps reveals its stark contradictions best in Nazi Germany.[47] A reenchantment of Nature in Nordic myth and new Aryan ritual produced V-2s, Auschwitz, ME-262s, and nuclear fission, while covering itself in fables of Teutonic warriors true to tribal *Blut und Boden*. Industrial fascism in Germany openly proclaimed itself to be *antimodern* and *future primitive*. It also condemned industrialism and the overpopulation of most other societies as it propounded a very peculiar vision of reinhabiting its self-proclaimed and historically denied bioregional Lebensraum. One should not assume that deep ecology will lead to a fascist outcome; yet the deep ecologists must demonstrate why their philosophy would not conclude in a similarly deformed fusion of modernity with premodernity.

Deep ecology's evocation of "Eastern spiritual process traditions," namely, Taoism and Buddhism, could be lifted out of its cultural context with little consideration of any sociocultural grounding. On one level, these religions may enable one to "relate to the process of becoming more mature, of awakening from illusion and delusion," and help "groups of people engaged in the 'real work' of cultivating their own ecological consciousness."[48] Yet, on another level, these traditions embrace an "organic unity" and, as the Tao bids, the quest for "production without possession, action without self-assertion, development without domination," which expresses their origins in traditional bureaucratic empires prior to or outside of the modern world system.[49]

In peasant villages bound by kinship-driven collectivism, limited agricultural productivity, and imperial bureaucratic oversight, as Max Weber notes, these religions served as a salvation ethic based on attaining perfect changelessness. In this "Eastern spiritual process tradition," "when such salvation is gained, the deep joy and tender, undifferentiated love characterizing such illumination provides the highest blessing possible in this existence, short of absorption into the eternal dreamless sleep of *Nirvana*, the only state in which no change occurs," which comes, in turn, by freeing oneself "from all personal ties to family and world, pursuing the goal of mystical illumination by fulfilling the injunctions relating to the correct path (*dharma*)."[50] For Weber, these ethics led down a path of world rejection by directing their followers to avoid individual self-realization. On the wheel of karma causality, each new incarnation of an individual creates or denies new chances for Nirvana attainment in the future. Any attempt at rational, purposive activity or efforts in the methodical control of life leads away from salvation.

Such ethical codes, if adhered to exactly, are ill suited to a purposive revival of ecological consciousness to revolutionize advanced industrialism. They are credos of world rejection and individual effacement (as might suit the spiritual needs of overburdened agricultural producers under oriental despotism) rather than a continuation of the individuation, personalism, nominalism, and existentialism Devall and Sessions endorse in primal cultures. To evoke such religious outlooks in postindustrial America, on one level, may promote maturity and forsaking consumerist illusions, while, on another level, providing an ineffectual opiate for the masses as their current material standard of living disappears in deep ecological reforms.

The Myth of Nature's Subjectivity

Deep ecology's ultimate value of self-realization claims to go "beyond the modern Western *self* which is defined as an isolated ego striving primarily for hedonistic gratification or for a narrow sense of individual salvation in this life or the next."[51] Real selfhood, it is claimed, derives from human unity with Nature, realizing our mature personhood and uniqueness with all other human and nonhuman forms of being. Humanity must be "naturalized"; that is, the "human self" is not an atomistic ego, but a species-being and a Nature-being as a self-in-Self, "where Self stands for organic wholeness."[52] Here, the essence of Nature, to a large extent, would appear to be a projection of an idealized humanity onto the natural world. Nature is "humanized" in a myth of subjectivity to change human behavior. The reanimation of Nature in deep ecology extends this selfhood to all natural entities—rocks, bacteria, trees, clouds, river systems, animals—and permits the realization of their inner essence.

As deep ecology depicts it, and as Georg Lukács would observe, Nature here

> refers to authentic humanity, the true essence of man liberated from the false, mechanizing forms of society: man as a perfected whole who inwardly has overcome, or is in the process of overcoming, the dichotomies of theory and practice, reason and the senses, form and content; man whose tendency to create his own forms does not imply an abstract rationalism which ignores concrete content; man for whom freedom and necessity are identical.[53]

Nature in this myth of subjectivity becomes for humanity the correct mediation of its acting that can generate a new, more just totality.

Deep ecologists, however, cannot really enter into an intersubjective discourse with rocks, rivers, or rhinos, despite John Muir's injunction to think like glaciers or mountains when confronting Nature. "The meditative deep questioning process" might allow humanity "an identification which goes beyond humanity to include the nonhuman world."[54] A hypostatization of self in human species being, whales, grizzlies, rain forests, mountains, rivers, and bacteria is no more than the individual's identification of his/her self with those particular aspects of Nature that express their peculiar human liberation. This ideological appropriation, in turn, is always (human) self-serving. One must ask, Is humanity naturalized in such self-realization or is Nature merely humanized to the degree that its components promote human "maturity and growth"?

This vision of self-realization appears to go beyond a modern Western notion of self tied to hedonistic gratification, but it does not transcend a narrow sense of individual salvation in this life or the next. Nature in deep ecology becomes humanity's transcendent identical subject-object. By projecting selfhood into Nature, humans are saved by finding their self-maturation and spiritual growth in it. These goals are found in one's life by in-dwelling psychically and physically in organic wholeness, as well as in the next life by recognizing that one may survive (physically in fact) within other humans, whales, grizzlies, rain forests, mountains, rivers, and bacteria or (psychically in faith) as an essential part of an organic whole. Nature, then, becomes ecosophical humanity's alienated self-understanding, partly reflected back to itself and selectively perceived as self-realization, rediscovered in selected biospheric processes.

Biocentrism as Soft Anthropocentrism

Biocentrism in deep ecology may simply be a spiritually refreshing form of anthropocentrism. Under certain conditions, Naess's claim that all organisms and entities in the ecosphere are equal in intrinsic worth and share an equal right to self-realization makes sense. If we deforest tropical Brazil or vent fluorocarbons into the atmosphere, we do deny rain forests and the ozone layer the right to existence. By fooling with Nature in this way, we also are foolishly harming ourselves. As Devall and Sessions claim, "there are no boundaries and everything is interrelated. But insofar as we perceive things as individual organisms or entities, the insight draws us to respect all human and nonhuman individuals in their own right as parts of the whole without feeling the need to set up hierarchies of species with humans at the top."[55]

Yet, Naess's "mutual predation" proviso belies this principle, because "in the process of living, all species use each other as food, shelter, etc.," which is "a biological fact of life." Biocentrism, then, if people are to survive, mystifies the workings of a "soft anthropocentrism." Even if people reduce their needs, and live with minimum rather than maximum impact on the earth, human interrelations with Nature still remain anthropocentric. Rocks and trees do not use humans—except perhaps for the molecules from decayed human bodies—for their survival or self-realization. Humans, however, do move, crush, and chip rocks as well as chop, carve, and burn trees to create shelter, tools, food, or medicines. Individual nonhuman entities or organisms are treated with less respect, equality, and rights by humans. Even if humans abdicate as masters of Nature, Nature still will feel the pressure of human hunters and gatherers.

Although humans could change some of their ways, such predation would be far from mutual. We might let sharks eat as many swimmers as they can find without reprisal or allow grizzlies to chow down on campers and livestock as their mode of self-realization. We could even solve the retirement crisis, prison overcrowding, or warehousing mental patients by staking the old, felons, and the insane out in tidal pools or on anthills. But, will we allow anthrax or cholera microbes to attain self-realization in wiping out sheep herds or human kindergartens? Will we continue to deny salmonella or botulism their equal rights when we process the dead carcasses of animals and plants that we eat? In the end, humans inevitably put themselves above other species and natural entities, as deep ecology accepts, "simply to live."

The norms of biocentrism, then, would reenchant Nature to make new spiritually denominated anthropocentric claims against the ecosphere. Polluted or abused natural settings cannot satisfy these spiritual demands; thus, deep ecology extends subjectivity to Nature as a means of limiting environmental abuse. As Devall and Sessions basically admit, the norm of biocentric equality would guarantee "an overriding vital human need for a healthy and high-quality natural environment for humans, if not for all life, with minimum intrusion of toxic waste, nuclear radiation from human enterprises, minimum acid rain and smog, and enough free flowing wilderness so humans can get in touch with their sources, the natural rhythms and flow of time and place."[56] Dressing up such human-centered appropriation of Nature in rituals or myths would make it more psychically rewarding or spiritually refreshing, but biocentric equality comes across as little more than "soft anthropocentrism." It seems doubtful that even a primal hunting and gathering society could consistently meet the strictures of a "hard biocentrism," much less an ecologizing postindustrial society.

The Modern Core of Ecological Subjectivity

Deep ecology's critique of the Enlightenment schema is neither as thorough nor as radical as its advocates claim. By citing new norms to constrain humanity's destruction of the ecosphere, deep ecologists aspire to overturn the Enlightenment schema underpinning advanced industrialism's instrumental rationality. In adopting examples they see in primal cultures, deep ecologists believe they can effectuate Nature's reenchantment, develop nondominating sciences, and launch a new ecological society by creating new forms of human selfhood.[57] Although deep ecology presents these goals as tantamount to the abolition of man's domineering power over Nature, it appears instead that human power would not be replaced by biocentric equality as much as it could be displaced by a silent anthropocentrism in this new human subjectivity.

The new philosophy of nature might seal "the death of man" in ecological functioning by supplanting a coercive set of human power relations with a new discipline of ethical surveillance (self-administered by the subject in Taoist meditation, Buddhist self-in-Self introspection, and mythic Amerindian purification rituals) to reconstitute human agency within natural subjectivity. The sites of power plainly would shift, because the disciplines of self and social understanding would be forced into new polarities of value and practice.[58] In constituting biospheric entities as subjects, humanity would become, following Aldo Leopold's paradoxical idealization, just "plain citizens" in an egalitarian biotic/geological/atmospheric community.[59] The strategies of ecosophy would shift human power over Nature (and humanity by implication) from external sovereign control in a Hobbesian sense to internal participative normalization with Nature in a new Foucauldian sense.

Much of the modern Enlightenment schema could survive these

transformations.[60] Enforcing harmony with Nature might be as destructive and domineering as attaining dominance over Nature. Deep ecology's construction of reenchantment, mature selfhood, and Nature bear the birthmarks of modernity in its reconceptualization of the postmodern as primal premodernity. In this regard, deep ecology's confrontation with technocratic industrialism mirrors Rousseau's confrontation with the Enlightenment.[61] The good person, or ecosophical people, should follow "the voice of Nature," not "the voice of Reason," which simply expresses instrumental strategies for satisfying corrupt social desires. As subjects of the dominant worldview, people disenchant the world and seek instrumental control in the false voice and language of Reason. Yet, these corrupting social forces interfere with people's sensing and following of their true natural sentiments. If we develop, as Naess, Devall, or Sessions claim, our intuition of Earth Wisdom, or attune our sentiments to Nature, then we might tap into new virtuous realms of true freedom. Even so, the pure voice of Nature, speaking through individual conscience in the language of virtue, expresses a very modern concern for individual self-realization of each unique quality to its utmost in each subject's being.

Although deep ecology casts this shift in subjectivity as a revival of primalism, it also might be a Rousseauian revitalization of self-expressive modernism. Nature here speaks of virtues and freedoms that are those of sovereign individuals, creating themselves by rescuing their selves from the corruptions of modernity in finding their personal freedom in it. Nature is a healing force or foundation of virtue, which, once recovered in the "real work" of self-realization, empowers persons to attain full self-expression. As with Rousseau's modern ethic of individual expression, each of us—humans, rocks, or rivers—has a nature to be revealed, expressed, realized in complete and equal self-fulfillment. Modern subjectivity is not so much overcome as it is made into an equal entitlement and guaranteed to everything in the ecosphere.

Deep ecology tends to "green" or "soften" the Enlightenment schema, but it does not overturn its workings. Primal pre-Enlightenment traditions are pared down to suit the particular needs of some postmodern intellectuals, who take from them only what they need to assail advanced industrialism's ecological abuse. Diverse types of reenchantment, myth, and ritual also are embraced without much thought to their operational interests or their probable legitimation of a new hyperinstrumental science. An ethic of self-realization is espoused, like Rousseau's *Émile*, that projects modern individual self-expression as an ontological quality of nature.[62] By the same token, the precept of biocentrism seems to occlude a soft anthropocentrism in issuing a license of "mutual predation" to Nature's most successful and destructive predator—people. Finally, deep ecology could function as a new strategy of power for normalizing new ecological subjects—human and nonhuman—in disciplines of self-effacing moral consciousness. As the new philosophy of nature, deep ecology could provide the essential discursive grid for a few ecosophical mandarins to interpret Nature and its deep ecological dictates to the unwilling many.

The Politics of Deep Ecology

Given these conceptual problems in the deep ecological program, how can deep ecology be situated politically within the existing system of power? In the last analysis, Naess, Devall, and Sessions respond to the environmental crisis with "an examination of the dominant worldview in our society, which has led directly to the continuing crisis of culture," and their practical political answers essentially present "an ecological, philosophical, spiritual approach for dealing with the crisis."[63] Any post-Marxian political critique must come to terms with this insight: "it is not the consciousness of men that determines their being, but, on the contrary, their social being that determines their consciousness."[64] In view of Marx's claim, what sort of social being or class position has determined the shape of deep ecological consciousness?

Chim Blea (aka Earth First!'s Dave Foreman) states: "Deep ecology has been developed by outdoorspersons—mountain climbers, backpackers, field biologists—with experience in observing natural phenomena and comes from the conservation/preservation movement," and it "seeks to develop a new paradigm, questions the essence of human civilization, fundamentally condemns human overpopulation and industrialism, is *anti-modern* and *future primitive*, bio-regional, reinhabitory, and resacralization."[65] Not surprisingly, deep ecologists look to these people and their practice for the discipline, rituals, and teaching of ecological consciousness. Taking care of a place, bringing an attitude of watchful attention to the environment, focusing on self in Nature, finding maturity and joy in natural being, and simply doing outdoor activities all are basic values shared by many outdoorspersons. If they are done "with the proper attitude," many personal leisure pursuits, "like fishing, hunting, surfing, sun bathing, kayaking, canoeing, sailing, mountain climbing, hang gliding, skiing, running, bicycling and birdwatching,"[66] are endorsed as a path to attain clear ecological awareness.

These modes of social existence do determine the consciousness and practical programs of the deep ecology movement, but they also expose a mystification. Many outdoor activities for finding the right ecological dharma are highly industrialized modes of corporate consumerist leisure. Free-form mountain climbing, fishing, running, or sunbathing do not demand high-tech equipment, but bicycling, surfing, hang gliding, skiing, hunting, or kayaking are among the most deeply entrenched bastions, as Edward Abbey decries, of "industrial tourism."[67] A self-contained industrial tourist, who brings all of his/her food, clothing, shelter, and equipment from the leisure industry or supermarket, can pretend to relate to Nature as a biospherical equal by finding self-realization on a sunny weekend. Many ordinary workers and farmers know and enjoy the outdoors on the right ecological terms, but deep ecology, as a philosophy for properly outfitted mountain climbers, backpackers, and field biologists, could also be mistaken for the ideology of white-collar intellectuals or professional-technical yuppies defending their environmental "positional goods."[68] Rock cliffs, back country powder, trout pools, forest trails, monster breakers, elk herds, class three river rapids, or bird preserves can be enjoyed more thoroughly by a few individuals to the extent that most others cannot enjoy them. It makes sense for deep ecologists to condemn human overpopulation or resacralize the bioregion they wish to enjoy. Unfortunately, nomadic grub eaters cannot produce high-tech composite surfboards, eighteen-speed bicycles, or sophisticated hang gliders. Who will make such goods or produce food while others seek self-realization and biocentric equality? The antimodern, future primitive condemnation of industrial human civilization by many deep ecologists is not really total, but its contradictory partialities are mystified in the social forms of life that generate this consciousness.

Beyond such mystifications of its class origins, the eight essential principles of deep ecology elaborated by Naess and Sessions are lacking as a practical political program.[69] First, the notions that (1) the well-being of human and nonhuman life has intrinsic value apart from human uses, that (2) the diversity of life forms contributes to these intrinsic values, and, finally, that (3) humans have no right to reduce this rich diversity except to satisfy vital needs are all important goals. Chim Blea, for example, argues that all things are equally valuable, from plants to people to clouds, but animals are no more crucial than plants or mammals than insects.[70] However, rare species and endangered individuals in rare species as they become endangered are more valuable than more abundant species and individuals of such species. So, if a deep ecologist was caught in a spring brushfire, would she be bound to save a rare California condor hatchling over a human child, because the former in some sense is much more valuable? Nature is believed to know best. Humanity should not assign good or evil labels to the cycles of suffering, pain, and death in Nature—people should just let it be.

Deep ecologists offer few criteria for practicing these precepts. If humans have no right to reduce the diversity of life, except to satisfy vital needs, then what are the standards for identifying vital needs? Suffering, pain, and death are defined as natural or inevitable. Can humans destroy viruses and bacteria to cure disease? Can humans eat all plants if game is meager or turn to any game animals if crops fail? Can humans reduce the diversity of a river basin with dams to control floods that let rivers or rain become what they are, and allow crops to displace natural vegetation and alter nitrogen fixation cycles in the existing topsoil? Nature has let California condors, the tiger, African elephants, and blue whales suffer and die in their ecological niches, allowing human beings or draft animals to supplant them. Should humanity act otherwise, or let it be?

Deep ecology should not necessarily have all the answers now, but it does need substantive criteria for arriving at answers in the future. Right now, however, all that deep ecology seems to offer are new symbolic rituals of sacralization instead of substantive rational criteria for choosing between alternatives. As Chim Blea suggests,

I thank pieces of wood I gather for my fire, I treat as sacred whatever I eat or use to fulfill vital needs whether it be animal, plant or mineral. . . . We as Deep Ecologists recognize the transcendence of the community over any individual, we should deal with all individuals— animal, plant, mineral, etc.—with whom we come into contact with compassion and bonhomie. Some we will use to fulfill vital needs, some will use us to fulfill their own vital needs, some (like burros in the Grand Canyon) we may need to kill to protect the integrity of the community, but all should be treated with respect and love.[71]

Here, again, there is a soft anthropocentrism at work in Nature sacralization. People must continue to cut and burn trees, kill and eat plants and animals, or isolate and kill germs to fulfill vital needs in humanly defined ecocommunities. As Alan Watts first noted, when humans want food, "cows do scream louder than carrots,"[72] but deep ecologists do not draw such distinctions or provide criteria for judging between cows and carrots, rocks and rivers, or people and ptomaine. After deep ecological training, a ritual prayer, the right attitude of respect, or compassionate loving gratitude will rationalize and legitimate anthropocentric actions. Serious abuses of Nature will probably lessen, but the human being still is "more equal" than other beings in deep ecology's Animal Farm.

Second, the idea that (4) human life must reduce its population to flourish itself as well as to promote nonhuman life, because (5) human interference in the nonhuman world is excessive and worsening presents many practical difficulties. These principles may be true, but who decides how to decrease human populations where, when, and why? In the developed world, human life sees the underdeveloped world overpopulating itself, forcing Nature into collapse, whereas in the underdeveloped world, more human life in new children promotes the continuation of their parents' individual human lives. And, excessive consumerism in the developed world is seen by the underdeveloped world as excessive, causing Nature to collapse. Most people will not voluntarily stop reproducing to protect themselves, much less the survival of Nature. Even so, if people did come into Earth Wisdom, reproduction is "natural." It is "letting Nature be" by finding one form of self-realization in new life. To make these principles actually work, vital human needs, such as food, shelter, clothing, health, or life itself, probably cannot be satisfied. A "hard biocentrism" versus a "soft anthropocentrism" would reduce the human population. But, existing technologies of comfort and security would have to be suspended or outlawed as famine, disease, the elements, or reinhabitory predators reduced human populations.

Third, principle (6) is the least developed and the most problematic: policies must change to transform the economic, ideological, and technological structures into something not like today's, which implicitly will fulfill the dictates of principles 1 through 6. Beyond precept (7) shifting human satisfaction to appreciate the quality of life over higher quantitative standards of living, deep ecology has no program. At best, it only offers the traditional solution of social anarchism, changing the self to change society with individual acts of will guided by correct conscience/consciousness in a pressing situation of necessity to save humanity. Devall and Sessions claim, "if we seek only personal redemption we could become solitary ecological saints among the masses of those we might classify as 'sinners' who continue to pollute. Change in persons requires a change in culture and vice versa. We cannot ignore the personal area nor the social, for our project is to enhance harmony with each other, the planet and ourselves."[73] This outlook leads to principle (8), namely, those who subscribe to these points have an obligation to directly or indirectly implement them. These feeble injunctions will not empower the minority tradition in ecological communities. There is not a concrete theory of the state, ideology, technology, or the economy here. Deep ecology fails to admit how people enthusiastically volunteer in the rape of Nature to enjoy corporate consumerism. Today's Michelob philosophy of "You Can Have It All," while "living on the edge" often *is* seen as spiritually more satisfying than Earth Wisdom. Basically, deep ecologists fail to ask or answer Lenin's question of advanced industrialism: "Who, Whom?"

Ultimately, deep ecology is "utopian ecologism." As a utopia, it presents some alluring moral visions of what might be; at the same time, it fails to outline practicable means for realizing these moral visions. Deep ecologists are caught in the trap of endorsing new visions for new "ecotopias," but they do not even have a practical program for future primitive reinhabitation or bioregional community building. Political action is displaced into the realm of ethical ideals, making it every individual's moral duty to change himself or herself in advancing cultural change. Without the opportunity to change collective activity—in the economy, ideology, technology, or polity—this personal moral regeneration might become only a quietistic, postmodern Taoism of finding the right path in an evil society. Naess, for example, suggests that his vision of deep ecology is virtually idiosyncratic; others are strongly enjoined to concoct their own ecological omelettes.[74] Devall and Sessions conclude that deep ecology stands for these ultimate values:

> Inward and outward direction, two aspects of the same process. We are not alone. We are part and parcel of the larger community, the land community. Each life in its own sense is heroic and connected. In the words of Bodhisativa, "No one is saved until we are all saved."
>
> This perspective encompasses all notions of saving anything, whether it be an endangered species, the community, or your own self. Each life is a heroic quest. It is a journey of the spirit during which we discover our purpose. We have only to embark, to set out in our own hearts, on this journey we began so long ago, to start on the "real work" of becoming real and of doing what is real. Nothing is labored, nothing forced.
>
> The process of developing maturity is simpler than many think. Like water flowing through the canyons, always yielding, always finding its way back, simple in means, rich in ends.[75]

The deep ecologists may claim these values as their final goals. However, such principles have little practical utility for staging an ecological revolution.

If the economy, ideology, and technology of corporate consumerism are to change, then one must ask: Who dominates whom? How? Why? Where? What really is to be done? Deep ecology does not address these questions or provide any accurate answers. Deep ecologists also dismiss the important contributions being made by reform environmentalism in the areas of alternative agriculture, alternative technology, or alternative architecture. Beyond engaging in nonviolent resistance while acting from deep principles "to touch the earth," there are no effective strategies for real change in deep ecology, save those of continuing the tactics of reform environmentalism with a new, deep, long-range attitude "to better public policy" through frugality, modesty, and restraint.

Justice and Deep Ecology

Deep ecology's acceptance of otherness in nonhuman life and inanimate entities in the ecosphere is an important contribution. Deep ecologists identify a new normative ethic of personal responsibility in caring for Nature that has basic merit. Yet, as political philosophy, deep ecology has failed thus far to demonstrate how it can be implemented anywhere today. Like many revolutionary programs, deep ecology lacks a theory of the transition. There are no practicable means for changing the everyday life of everyone in the stage of advanced industrialism into an ecotopian community without tremendous costs. Many would agree with Snyder that "we must change the very foundations of our society and our minds. Nothing short of total transformation will do much good."[76] But, how does the United States with 250 million people, living because of the imports and exports of transnational corporate capitalism in and out of huge metroplexes, reinhabit its bioregions such that "the human population lives harmoniously and dynamically by employing a sophisticated and unobtrusive technology in a world environment which is 'left natural'"?[77] Current world urbanism assumes an obtrusive technology that renders the organic into the inorganic. What happens to Los Angeles, Chicago, New York? Where do these millions go and what will they do? If their corporate agricultural or municipal service supports are cut simply to return the L.A. Basin, Lake Michigan's South Shores, and Manhattan to Nature, then Nature does know best how to cope—these immense human populations will suffer and/or die.

Deep ecological justice is postdistributional. It defines away distribution systems with human norms of fairness or equality as the apparatus of corrupt technoindustrial society. By calling for biospherical egalitarianism, deep ecology extends the right of life, liberty, and the pursuit of happiness (as the freedom of self-realization) to nonhuman life and inanimate entities so that humans, for the first time, can truly enjoy their rights to life, liberty, and the pursuit of happiness in emancipated Nature. Justice is made into an attribute of all-selves-in-Self working toward their peculiar self-realization.[78] Therefore, humans must alter their hitherto anthropocentric modes of existence, out of the new sense of "fairness" to otherness and other humans growing from ecosophical consciousness, to promote this new biocentric justice.

This quest for "natural unity in process" or "biospherical egalitarianism" bears many inner contradictions. The soft anthropocentrism of deep ecology, which favors the nature-regarding interests of humans, spreads intrinsic value evenly across the ecosphere in principle as it continues to overvalue certain humans over other humans, animals, plants, and inanimate entities. As Fox argues, "the only universe where value is spread evenly across the field is a dead universe."[79] Deep ecologists, in stressing first principles, really do not develop an adequate theory of justice or a workable system of ethics. While acknowledging that killing and suffering are natural or necessary, they advance no criteria for deciding between alternatives when such "natural acts" become necessary. As the next chapter on Earth First! shows, deep ecology provides few guideposts on this rougher ground. Until it comes to grips with these contradictions, deep ecology must be held suspect as a political philosophy.

2

Ecological Politics and Local Struggles: Earth First! as an Environmental Resistance Movement

This chapter analyzes the operations of Earth First! as an environmental resistance movement, one that has been influenced significantly by deep ecology thinkers. Working on the local level in the United States as well as Australia, Canada, Denmark, Germany, India, Mexico, Poland, and Great Britain, many Earth First! activists have proven to be radical practitioners of ecological direct action to protect wilderness areas and preserve biodiversity.[1] At the same time, the federal government, national broadcasting networks, and local media outlets in the United States have expended a great deal of time and energy portraying Earth First! as a new kind of terrorist or criminal group.

Discourses of "terrorism" or "criminality," however, provide an inadequate interpretive frame for understanding the particularistic, biocentric, and localistic strategies of Earth First! ecological activism.[2] This chapter instead reexamines Earth First! as a new social movement, whose project expresses alternative philosophies of "life world defense," "environmental liberation," or "ecological resistance." Working outside of tactical conventions followed by many other environmentalist organizations, Earth First! provides evidence of how often today's new social movements diverge from previous modern popular protest movements.[3]

By situating Earth First! in contemporary political discourses about new social movements, this chapter represents it as an example of how some of these groups are developing innovative new strategies of political resistance. First, it reevaluates Earth First!'s program for opposing today's industrial society, stressing the new departures in its organizational structures and operational strategies from the prevailing reformist

28

approaches of defending wilderness and biodiversity. Second, it discusses the major points of its political program, including its global perspective on mobilizing new groups to engage in local environmental defense. Third, it considers how localistic strategies and sites of struggle, when given extensive attention by the mass media and state authorities, provide a basis for a transnational opposition to contemporary corporate capitalism's rationalization of everyday life. Finally, it indicates why Earth First!'s political structures and ecological philosophies might exemplify some of the workings of contemporary new social movements.

The Cultural Logic of New Social Movements

After seven decades of struggle against socialist and/or communist opposition groups, the established cultures of governance in the United States almost always have displayed befuddlement when confronted by new, nonsocialistic forms of resistance. The experience of one of Earth First!'s founders, Dave Foreman, at an antilogging protest site in Oregon demonstrates this confusion. "One day," at a Forest Service road-building project on national forest land outside Medford, Oregon,

> only Foreman and a guy in a wheelchair showed up to stand in front of the machinery. The truck driver kept coming slowly forward, bumped him, and then Foreman grabbed the bumper. He was dragged about a hundred yards when the driver stopped, jumped out and denounced him as a Red. Foreman replied that he was [as, indeed, he is and always has been] a registered Republican.[4]

Like most new social movements, Earth First! is post-Marxist in its politics and anti-Marxist in its philosophical program. In fact, the activity of most new social movements, such as Earth First! and its various allied ecological resistance groups,

> explode[s] the myth of a single set of laws governing all historical development: the feminist, peace, ecology, and radical cultural movements, for example, are both too pluralistic and autonomous to be neatly subsumed under a unifying logic of capitalist production, even if their growth is powerfully influenced by it. These popular revolts express a wide diversity of interests, priorities, goals, and ideologies; theoretical efforts to force it into a single pattern will only obscure its

essence and, hence obscure the very dynamics of change in advanced capitalism.[5]

Unlike most "old social movements," which were inspired by Marxian, liberal, or socialist programs for redistributing material resources, "new" social movements often have adopted agendas aimed mainly at redefining cultural identities as a form of political or economic resistance.

Earth First! pursues goals that are *not* mainly economic; instead it consciously struggles "over the power to socially construct new identities, to create democratic spaces for autonomous social action, and to reinterpret norms and reshape institutions."[6] This dovetails with Jürgen Habermas's reading of new social movements as fragmentary efforts made by groups within an embattled but still existing civil society, struggling to defend independent identities and cultural autonomy outside of the institutionalized structures of power and exchange in contemporary corporate capitalism. Such groups fight against "the 'productivistic core of performance' in late capitalist societies" by advancing alternative cultural visions to organize their everyday life world.[7] Operating from within gaps in the already highly rationalized life world of advanced capitalism, these new activists "seek to *stem* or block the formal, organized spheres of action in favor of communicative structures; they do not seek to conquer new territory."[8] Habermas argues that these types of groups directly oppose "the profit-oriented instrumentalization of professional labor, the market-dependent mobilization of labor, and the extension of competitiveness and performance pressure"[9] into the spheres of everyday life. Rather than relying on a Marxian class conflict model of social antagonism, Habermas sees other cultural and institutional contradictions growing from the opposition of "civil society" and the "state." As elements of a collapsing "civil society," new social movements function as resistance and liberation movements, fighting the "state" and/or the "market," and their colonization of the everyday life world with the logics of power and exchange through bureaucratic or commercialistic codes of instrumental rationality. "In short," Habermas claims, "the new conflicts are not sparked by *problems of distribution*, but concern the *grammar of forms of life*."[10] Habermas positions the new social movements in his model on "the periphery" as "outsider groups" struggling against "core" institutions and "insider elites" over the scope and content of the formal rationalization taking place in the social life world.

Taking this approach further, Joachim Hirsch maintains that "technocratic modernizers" in the "Fordist security state," who dominate the state under advanced capitalism, have become "insensitive and unresponsive to social interests and problems. Thus, larger parts of the population no longer feel truly represented. Hence social conflicts and problems unfold outside the bureaucratic sphere of control and perception."[11] The advancing rationalization of the ordinary life world splits corporate capitalist societies into two divergent fragments, which are based, in turn, on two interconnected but distinctive political economies: a *modern productivist sector* and a *marginal nonproductivist sector*. As this division deepens,

> non-productivist interests—like those in a healthy environment or in natural ecology—are marginalized within and across individual people. An example would be the justification for the destruction of the cities and of the natural environment by pointing to secure employment (as is the case in the nuclear and automobile industries). Here lies the material basis for the recent discussion of the so-called "change of values." Because of this development, social conflicts still result from the content of capitalist exploitation, yet they do not manifest themselves along traditional class lines. Nor can such conflicts find expression within the system of political apparatuses, because they are structurally excluded.[12]

Along this divide between productivist and nonproductivist interests, the extrainstitutional protests of new social movements provide a new mode of opposition. In advanced capitalist nations, resistance is expressed in antibureaucratic, antitechnocratic, anticorporate modes of cultural and political mass mobilization. Hirsch concludes:

> The rigid and opaque structure of the political system promotes the rise of these movements, which try to articulate and accomplish neglected needs and interests. As they do not correspond to the established system's notion of functional logic, they necessarily (and frequently without intention) are in opposition to it. These "new social movements" find expression in several citizens' initiatives, in the ecol-

ogy movement, as well as in spontaneous strikes or the occupation of factories.[13]

Many new social movements show that there are alternative visions of modernity predicated on developing new economic systems, social institutions, and cultural values that might preserve more autonomous spheres of personal identity and social action from the onslaught of instrumental reason. They present alternative grammars for the basic forms of life, which might stress nonproductivist interests, discursive decision-making procedures, and new modes of modernizing change that would empower clients, consumers, and citizens as their innovative tactics resist the colonizing influences of state/corporate technocratic modernizers.[14] Earth First! is interesting inasmuch as its "interpretations of the life world" draw hard boundaries around the further commodification and rationalization of Nature. It aims, in one sense, to defend whole new life worlds beyond human society—natural wilderness and biodiverse ecosystems—from all, or at least all but the most primitive and least intrusive, human interventions.

The Origins and Operations of Earth First!

As a new social movement fighting to redefine cultural identity by protecting the rights of wildlife and wild places in Nature, Earth First! developed a loose, weblike style of organization. It must be remembered that the loosely organized Earth First! groups are not the entire radical environmental movement, nor the whole deep ecology movement. On this point, as Foreman once argued, "Earth First! is not part of, nor are we, the reform environmental movement, the animal rights movement, the anarchist movement, the peace movement, the social justice movement, the anti-nuclear movement, the non-violence movement, the Rainbow Tribe, the neo-Pagan movement, the native rights movement, the Green movement, or the Left."[15]

Although there are areas of overlap and affinity between all of these diverse groups and elements of Earth First!, Foreman asserts that Earth First! has very unconventional roots. Most critically, this movement did not emerge "from the anarchist movement, nor from the Left. Earth First! came very directly out of the public lands conservation movement—out of the Sierra Club, Friends of the Earth and the Wilderness Society. It is public lands issues and wilderness that have been central to us from our formation."[16] Foreman's previous career with the Wilderness Society as a lobbyist, as well as the fairly nonideological background of Earth First!'s other founders, only serve to further underscore this point.[17] Particularly during the years in which Foreman was Earth First!'s nominal "head" or "spokesperson," these origins in conservationist or game management thinking also were what gave Earth First! much of what many regard as a somewhat chilling social philosophy.[18] Wilderness-centered Earth First!ers tend to look at human beings ecologically, or as one more "natural population" that has exceeded the carrying capacity of its range; hence, like rabbits, algae, deer, or locusts in similar circumstances, there must be a catastrophic crash or mass die-off to reequilibrate networks of ecological exchange on the range. This reduction of human beings to nothing more than an animal population, the assessment of all life forms as being equal in value to maintaining the range, and the acceptance of major extinctions among humans to protect plants, animals, and landscapes all illustrate the difficulties in turning game management techniques into social theory. After Foreman distanced himself from Earth First! during 1991–92 over new members using more "sociocentric," or traditional leftist, tactics and rhetorics, these wilderness-centered, preservationist philosophies have been downplayed. Nonetheless, Earth First!'s political agendas still follow, for better or worse, a radical preservationist approach to wilderness. Its wilderness program clearly grows out of game management science; it does not derive from its politics.[19]

Earth First! is not a large movement. Its activist core is probably no more than five hundred, while its less active supporters in the United States and Canada could be as many as fifty thousand. Another ten thousand supporters most likely exist around the world outside of North America. The *Earth First! Journal*, which is the most heavily subscribed to publication among all Earth First! groups, has a readership estimated to be as high as fifteen thousand.[20] The movement was founded in April 1980 by Dave Foreman along with Mike Roselle, Bart Koehler, Howie Wolke, and Ron Kezar on a trip to the Pinacate Desert in Sonora. Although it started out as a small, shoestring operation in Tucson, Arizona, Earth First! has grown substantially since then. In terms of its individual membership and number of affiliated groups, it now is a bigger, shoestring operation spread across many states and several foreign countries. By the mid-1990s, there were chapters operating in thirty-six states from every region of the United States.[21]

Although its members can meet at large group conferences, which are staged annually as the "Round River Rendezvous," with an open invitation for all interested persons to join in the proceedings, no bureaucratic control is exerted over the various local groups and coordinators.[22] Instead, each group and every coordinator essentially are free to determine their own agendas and programs in accord with their particular interpretations of the overall goals of Earth First! and local ecodefense needs. Intrinsically democratic and largely anarchical, Earth First! runs through a loose, weblike rather than rigid, pyramidal form of organization, which is inspired by both the member groups' persistent lack of extensive material resources and their philosophical desire to emulate more "primitive" societies' institutions. Never very tightly organized, Earth First! started fragmenting over strategies and philosophies during 1989–90. In May 1989, Dave Foreman and three other Earth First! members were arrested for plotting the destruction of various nuclear power facilities in the Southwest and a ski resort in Arizona.[23] And, in May 1990, Judi Bari and Darryl Cherney, two California Earth First! activists involved in planning the "Redwood Summer" campaigns against redwood timbering in Northern California, were injured by a car bomb as they drove through Oakland, California.[24]

The reaction of many Earth First!ers to these events accentuated the divisions that had been building in the movement during the 1980s. On the one side, an alleged "old boy network" of Earth First! founders continued pushing for preservationist, biocentric agendas apart from explicitly social and political concerns. On the other side, a newer collection of feminists, anarchists, revolutionaries, and organizers stressed for more openly political and overtly activist approaches to advancing Earth First! philosophies beyond tree spiking, banner hanging, or tree-sitting tactics used in the movement's early days. This shift was marked when the Montana editorial collective took over the *Earth First!* journal in 1990–91:

> While the values of the wild are at the heart of our project, this paper will not be primarily about wilderness and biodiversity. It will be about *defending* wilderness and biodiversity. . . . It will have discussions of various wilderness defense tactics, both legal and illegal, ranging from outrageous actions to the equally important mundane daily work of protecting wild things. It will include other discussions relevant to activists, concerning such topics as police surveillance and harassment, general organizing, and low impact living.[25]

Following this notion of *defending* wilderness and biodiversity, Earth First! activists began thinking about organizing timber workers, mobilizing mass demonstrations, and admitting to revolutionary aspirations instead of quietly and covertly sticking to hit-and-run monkey wrenching tactics.

The splits within Earth First! were formalized in 1991 when Dave Foreman and a number of his Earth First! supporters started their own journal, *Wild Earth*.[26] As advocates of biological diversity, Foreman presented himself and his sympathizers as striking out anew:

> *Wild Earth* is being launched to encourage this new approach to wilderness preservation. Our magazine exists as a forum for the serious discussion of the ideas and methods of *ecological preservation*. We are here to help translate the theories and information of Conservation Biology into grassroots preservation activism. We are here to help all groups and individuals to protect biological diversity. In doing that, we will consciously be advocates for non-human nature. We will speak for wolf, Orca, Gila Monster, Saguaro.[27]

Foreman and his fellow preservationists present themselves as returning to Earth First!'s original biocentric engagement of protecting nature. He argues: "We are conservationists. We believe in wilderness for its own sake. With John Muir, we are on the side of the bears in the war industrial humans have declared against wild nature."[28] In contrast, Foreman saw Earth First! turning in the 1990s to more sociocentric modes of ecological resistance, ranging from ecofeminist nature worship to radical anarchism to New Left anticapitalism. While Foreman sees himself launching "the new conservation movement," Earth First! continues its localistic environmental resistance actions, only now with often more admittedly "sociocentric" or apparently "anthropocentric" anarchist tinges.[29]

Organizationally, however, Earth First! still likens itself to visions it has constructed of a Native American "Indian tribe," and it is the vague images of "Indian tribes" held by its original founders—as they gathered them from industrial society's reigning discourses on Native American tribal organization—that often still serve as its leadership's institutional model. As John Davis noted, "Earth First! was established as a tribe. EF! never had *official* leaders. The leaders have been those who organized local groups and campaigns and initiated actions."[30] Although it may have been established by its founders as a tribe, Earth First! also has developed, using these organizational designs, into a global movement of highly motivated activists. "Instead of a central bureaucracy," Davis claims, "we have autonomous entities—local and regional EF! groups, task forces, roving individuals—who essentially shape their own agendas. This places tremendous responsibilities on individuals in EF! An Earth First!er is not simply a person who sends in an annual fee for membership, as a Sierra Club member may be; an EF!er is one who actively defends the planet."[31] Foreman explains that being part of the Earth First! tribe entails becoming an "ecowarrior," who willingly fights for the primacy of the Earth.

"Warrioring" is not necessarily a violent vocation. Instead, being an Earth First! warrior implies accepting a morally inspired confession of civil disobedience. As Foreman claims, "It doesn't have a violent connotation to it. It means that I am dedicated."[32] Rejecting models of bureaucratic operation drawn from the military, corporations, or states, Foreman saw Earth First! as a loose, antihierarchical assembly of these "warriors" fighting for biodiversity in the wilderness. "When you take on the organizational structure of the corporate state," he maintains, "you also tend to take on its ideology, and that if we were going to stay true to our biocentric principles, the only organization structure to emulate would be the hunter-gatherer tribe, which means no formal leadership, no hierarchy, and no formal membership."[33]

Earth First! also avoids relying on the formal institutions of conventional campaigning, fund-raising, public interest litigating, and pressure group lobbying used by what it calls "mainstream environmentalism."

Such mainstream groups have won only limited victories in gaining partial protections for particular areas, species, or ecosystems.[34] Yet, they have not stopped, or even lessened, the continuing destruction of wilderness areas, plants, and animals in North America. Outside of North America, the success of tactics used by mainstream environmentalism is even more constrained. As these conventional strategies fail, Earth First! activists engage in direct action, ranging from blockading roads, chaining themselves to trees, locking themselves to logging equipment, sitting in trees slated for cutting, climbing tall man-made structures to unfurl protest banners to engage in "monkey wrenching."

Monkey wrenching is essentially small-scale economic and technological sabotage directed against the complex of abstract machines used by businesses or government in destroying the wilderness. It has involved spiking trees (to make cutting and milling them dangerous), burning or disabling heavy earthmoving and timbering equipment, destroying trails and roads, slicing or spiking tires, cutting down billboards, and dumping trash on the property of its industrial or corporate producers. These tactics, of course, are the source of Earth First!ers' public notoriety as well as the major motive behind their harassment by the FBI, U.S. Forest Service, Bureau of Land Management, and local police forces. Yet, these tactics are not all used by every group. Some local groups even have officially renounced the more controversial and dangerous ones, such as tree spiking, as being counterproductive.[35] Because, however, Earth First! is not a bureaucratic organization with official membership criteria or certification procedures, these techniques of ecosabotage continue to be used by Earth First!ers and other freelance ecosaboteurs alike.[36]

The larger Earth First! movement provides at best only an aesthetics of action or a new narrative of Nature, but it is not an ideological monolith with its own credentializing administrative regime. Within its aesthetics of action, it enables individuals with sensitivity to Nature to define and describe their actions through new images and values tied to their new narratives of Nature philosophy. The local groups' reliance on individual conscience and commitment makes it difficult to call it a pressure group or political party in a conventional sense. Rather, it is an excellent example of how issue group politics coupled with direct action can produce a resilient resistance against environmental degradation. Mobilized by immediate concerns, guided by an evolving discourse of Nature and its worth, linked together by the *Earth First!* journal, occasional conferences, and personal contacts, Earth First! incorporates all the organizational possibilities of resistance in postmodern informational societies in one large but loose associational web.

To propagate its general principles of action, Earth First! publishes its own newspaper, *Earth First! The Radical Environmental Journal.* It is not, however, an ideological guidebook, specifying how and why activists should operate to be legitimately part of the movement. Similarly, it does not tout an official line to be transmitted at all levels from top to bottom in order to specify group ideologies and policies. The better-known figures in the movement are frequent contributors, but all tend to present their stories *as* "their stories" rather than as "the ideology" based on a certain command of special knowledge demanding that all implement it in everyday life or else. In fact, there is considerable controversy and disagreement in the pages of the journal, showing that it essentially serves as a bulletin board or poster session for various local efforts to get a larger hearing for their struggles.[37] In this role, one can read self-produced descriptions of different local chapters pursuing their Nature protection agendas. This coverage, however, illustrates the geographical diffusion of the movement and deepens popular support as it exposes potential new members to the general techniques of ecoraiding. Finally, like the publications of so many other groups in advanced capitalist countries, the journal also serves as "a marketplace." To raise money, each issue also has a "trinkets" section, hawking books, records, tapes, and videos on the general Earth First! approach to Nature, along with T-shirts, bumper stickers, and buttons inscribed with Nature-friendly agendas. Such commodified expressions of the memberships' political attitudes are simply part and parcel of doing mass politics in a market society where everyone increasingly purchases pieces of their personal identities in small boutiques of commodified community. Thus, far from being the ecological movement's functional equivalent of *Pravda*, each issue of *The Earth First! Journal* is much more like the radical environmentalist's *Car and Driver*, providing an open catalog of the essential ideas and symbolic artifacts needed to participate in and identify with a deep ecological discourse of ecological activism.

The Philosophical Program of Earth First!

Earth First! has drawn intellectual inspiration from many different sources. As what Dave Foreman has called "reluctant radicals," Earth First!ers ground their philosophical framework on the nature philosophy of influential figures from the American conservation movement. *The Earth First! Journal*, for example, identifies with "the tradition of John Muir, Will Dilg, Rosalie Edge, Aldo Leopold, Bob Marshall, Sigurd Olson, Rachel Carson, Howard Zahniser, Paul Schaefer, Marjorie Stoneman Douglas, Hazel Wood, Gertrude Blom, David Brower, Chico Mendez, Jose Lutzenberger and other radical conservationists."[38] However, in acknowledgment of Edward Abbey's inspiring visions of "monkey wrenching" in the Four Corners region of the American Southwest, *Earth First! Journal* recognizes that all Earth First!ers "also follow in the steps of some less public figures: Bonnie Abzug, Doc Sarvis, Seldom Seen Smith, Hayduke . . ."[39] Playing off of Edward Abbey's novels such as *The Monkey Wrench Gang* and *Hayduke Lives!* in which these fictional characters engage in ecologically minded sabotage, such as cutting down billboards, toppling power transmission towers, or blowing up dams, the intellectual outline of "monkey wrenching" encourages Earth First! activists and sympathizers to imitate Abbey's characters from the novels in hundreds of their own do-it-yourself acts of ecotage.[40] To make these tactics more uniform, Dave Foreman's own book, *Ecodefense: A Field Guide to Monkey Wrenching*, has articulated some methods for safely and successfully engaging in monkey wrenching in the real world.

Although the current *Earth First!* journal editorial group sees more differences between itself and deep ecology, when the journal was under Foreman's leadership, Earth First! largely accepted many of deep ecology's teachings in its general program: "We say that the ideas and manifestations of industrial civilization are anti-earth, anti-woman and anti-liberty. We are working to develop a new biocentric paradigm based on the intrinsic value of all things: Deep Ecology."[41] Deep Ecology, then, as it was first espoused by the philosopher Arne Naess and later popularized by Bill Devall and George Sessions, is plainly the other most important intellectual influence felt in the Earth First! movement.[42]

John Davis, a former editor of the *Earth First!* journal and now editor of *Wild Earth*, asserts that Earth First! represented a philosophy of ecological activism that is closely tied to biocentrism and deep ecology.

> EF! means recognizing that the planet and all its life forms have value (or dignity, or worth, or elan vital, or deoxyribonucleic acid or whatever it is that gives entities their reason to be) irrespective of their utility for humans. EF! means living in accordance with biocentrism—the principle that all natural life is equally central from the standpoint of the planet. This is diametrically opposed to anthropocentrism—the predominant worldview in human society, at least in the (over)developed nations. Anthropocentrism is the view that humans are the measure of all things, that things have value only insofar as they serve human ends.[43]

Although Earth First! echoes many of deep ecology's assumptions about nature, economy, and society, Davis also claimed, on the other hand, that "EF! is distinguishable from other elements of deep ecology by its emphasis on direct action and by its Luddite tendencies. Also some EF!ers don't like the term 'deep ecology' (saying it sounds like the study of benthic organisms) and simply speak of biocentrism (as advocated by Aldo Leopold long before Arne Naess coined the term 'deep ecology')."[44] Earth First!ers' "neo-Luddite" tendencies are, in fact, their most distinguishing characteristic. When it comes to complex modern technologies, "EF!ers question—if not outright condemn—technology, and some EF!ers deal with technological machines the way the Luddites (of 18th Century England) did—they bash them, or otherwise sabotage them so as to prevent their replacement of living entities. In contrast, some other deep ecologists look to appropriate technology—an oxymoron, many EF!ers say—to reverse the growing ecocatastrophe."[45]

Earth First! envisions its program as being substantively "posthumanist" or "biocentric."[46] It, like deep ecology, somewhat simplistically lumps most of Judaism, Christianity, Islam, capitalism, Marxism, scientism, communism, and secular humanism into the same threatening force—or what it labels "humanism." Choosing, somewhat strangely, Gifford Pinchot—the bureaucratic architect and first director of the U.S. Forest Service in the early 1900s—as the archetypical voice of humanism, Foreman sees all of these traditions as allied together in "the dominant philosophy of our time," which holds—in Pinchot's pithy quip—that "there are only two things in the world—human beings and natural resources."[47] This kind of reasoning can be defined even more explicitly. The "dominant philosophy of our time" in Earth First!ers' analysis is a mix of humanism, materialism, nationalism, and rationalism.[48] Resisting these principles provides the inspiration for Earth First!'s call of "Back to the Pleistocene" in reorganizing modern society. Adhering to this philosophy, while arguing for a capitalist versus a communist solution to the ecological crisis, to Foreman is silly: "EF!ers are not left or right; we are not even in front. Earth First! should not be in the political struggle between humanistic sects at all. We're involved in a wholly different game."[49]

This posthumanist attitude also presumes a biocentric stance against alleged anthropocentrism of humanism. Earth First!ers argue that "all things possess intrinsic value or inherent worth. Things have value and live for their own sake. Other beings (both animal and plant) and even so-called 'inanimate' objects such as rivers and mountains are not placed here for the convenience of human beings. The whole concept of 'resources' is denied by this philosophy."[50] The most basic concern of Earth First! is preserving wilderness from all human interference to maintain the range for all species. This issue is the crux of its "no compromise" strategies for liberating Nature from productivism: "Wilderness is the real world; it is our cities, our computers, our airplanes . . . our global business civilization which is artificial and transient. The preservation of wildness and native diversity is the most important issue. Issues affecting only humans pale into insignificance."[51] Biodiversity is stressed by EF!ers because of the urgent threat to many species of animals and plants—these forms of real life in the world *are* the real life forms that humans ought to care for in reconstructing their new social understanding of the life world. As Foreman maintains, up to one-third of all living species could be lost in the next two decades: "this is the greatest wave of species extinction in the 3.5 billion year history of planetary evolution—greater than the end of the Cretaceous—due to multinational greed."[52]

Foreman reminds Earth First!ers that the movement is called "Earth First!" not "People First!" Consequently, "in everything we do, the primary consideration should be for the long-term health and native diversity of Earth. After that, we can consider the welfare of humans. We should be kind, compassionate and caring with other people, but Earth comes first."[53] In turn, he argues that "a refusal to recognize the need to lower human population over the long run clearly defines one as a humanist and places them outside the bounds of Earth First!"[54] Foreman also holds that "an individual human life is not the most important thing in the world. An individual human life has no more intrinsic value than an individual Grizzly Bear life (indeed, some of us would argue that an individual human life has less because there are far fewer Grizzly Bears)."[55] Human famine and suffering in Africa are perhaps unfortunate, but in Earth First! the destruction of other species, wilderness, and natural habitats is viewed as being even more unfortunate. The ultimate logic of these statements—that if humans are *the* problem, then killing most of them would be *the* solution—is never really articulated. Foreman falls back on his game management theories as he reminds Earth First!ers that

> human beings are primates, mammals, vertebrates, animals. EF!ers recognize their animalness; we are not devotees of some Teilhardian New Age eco-la-la that says we must transcend our base animal nature and take charge of our evolution in order to become higher, moral beings. . . . We are in a struggle against the modern compulsion to become dull, passionless androids. We do not live sanitary, logical lives; we smell, taste, see, hear and feel Earth; we live with gusto. We are animal.[56]

This exaltation in animalness extends into engaging directly and frequently in real political action. Seeing many environmental groups as nothing more than debating societies, Foreman maintains that actions can set the finer points of Earth First! philosophies; that is, "we will never figure it all out, we will never be able to plan any campaign in complete detail, none of us will ever transcend a polluting lifestyle—but we can act. We can act with courage, with determination, with all the deliberateness we can muster, with love for things wild and free. . . . We have a job to do."[57] At the same time, Earth First! advocates that its members always practice what they preach. The existing structures of everyday life make living one's life in a truly ecological fashion remarkably difficult; nonetheless, "we are not able to achieve a true deep ecology lifestyle but it is the responsibility of each of us to begin to move in that direction. There are trade-offs. . . . We need to be aware of these trade-offs, and to make the best possible effort to limit our impact."[58]

The awareness of everyone's responsibility for the environmental crisis moves Earth First! to pardon no one for continual complicity in nature's destruction, including native peoples, women, racial minorities, or ordinary workers. In Foreman's view,

> the industrial workers, by and large, share the blame for the destruction of the natural world. They may be slaves of the big money boys, but they are generally happy, willing slaves who share the world view of their masters that Earth is a smorgasbord of resources for the taking. Indeed, sometimes it is the Hardy Swain, the sturdy yeoman from the bumpkin proletariat who holds the most violent and destructive attitudes towards the natural world (and towards those who would defend it). They are victims of an unjust economic system, yes, but that should not absolve them of what they do.[59]

This sort of language has not won Earth First! many friends among mainstream liberal or progressive political groups, even though the latest approach of Earth First!ers is to reach out to these groups for allies.

Foreman admits that "it hurts to be dismissed by the arbiters of opinion as 'nuts,' 'terrorists,' 'wackos,' or 'extremists.' But we are not crazy; we happen to be sane humans in an insane human society in a sane natural world."[60] With regard to corporate managers and government bureaucrats, EF!ers' judgments are starkly clear. By and large, "they are madmen destroying everything pure and beautiful. Why should we have any desire to 'reason' with them? We do not share the same world view or values."[61] Humanism relies far too much on formal rationality, but in Foreman's opinion

> rationality is just a fine and useful tool, but it is just that—a tool, one way of analyzing matters. Equally valid, perhaps ultimately more so, is intuitive, instinctive awareness. We can become more cognizant of ultimate truths sitting quietly in the wild than by reading books. Reading books, engaging in logical discourse, compiling facts and fig-

ures are necessary and important, but they are not the only ways to comprehend the world and our lives.[62]

Edward Abbey seconds this emotive faith in raw action: "Writing, reading, thinking are of value only when combined with effective action. . . . sentiment without action is the ruin of the soul. One brave deed is worth a hundred books, a thousand theories, a million words."[63]

Appeals to be "reasonable" and "responsible" in order to work out acceptable compromises for dealing with the environmental crisis are seen by Earth First! as one of the current regime's more potent counterpunches to radical resistance. Indeed, Foreman argues that becoming "reasonable" is the first step toward complete co-optation by the bureaucratic machines allied within state and corporate power.

> The American system is very effective at co-opting and moderating dissidents by giving them attention and then encouraging them to be "reasonable" so their ideas will be taken more seriously. Appearing on the evening news, on the front page of the newspaper, in a national magazine—all of these are methods the establishment uses to entice one to share their worldview and to enter the negotiating room to compromise. The actions of Earth First!—both the bold and the comic—have gained attention. If they are to have results, we must resist the siren's offer of credibility, legitimacy and a share in the decision-making. We are thwarting the system, not reforming it.[64]

Thwarting the system, in turn, demands that some monkey wrenching be done. For Earth First!ers, there is general acceptance among their many local groups that "properly done monkey wrenching is a legitimate tool for defense of the wild by some individuals" that clearly sets Earth First! "apart from other groups" and "helps define the specificity of being an Earth First!er."[65]

Earth First! is explicit about listing *how* wilderness is being destroyed as well as *who* is exploiting this destruction for profit. Each particular practice, in turn, is fair game as a potential target for Earth First! monkey wrenching and civil disobedience. Foreman identifies sixteen distinct activities that all are destructive of wilderness areas, including road building, logging, grazing, energy extraction, dams and flood control projects, power lines and pipelines, "slob hunting" and "slob fishing," wildlife management, introduction of exotic new species, destroying native species, off-road vehicle traffic, wilderness recreation, and "industrial tourism."[66] In the lower forty-eight American states, wilderness areas have become so developed and divided that no point is more than twenty-four miles from a road and many points are within ten miles of a road. Once roads are cut, "the army of wilderness destruction travels by road," including the practitioners of all of the sixteen most destructive activities identified by Foreman as being worthy of popular counterstrikes.[67]

Earth First! claims that the public guardians of wilderness in the federal and state governments are, in fact, totally corrupt compradors of corporate commerce. The Bureau of Land Management, which is charged with managing 180 million acres in the lower forty-eight states and mostly in the West, is seen as an agent of ranching, mining, and real estate interests as it develops annually more than a million acres. The U.S. Fish and Wildlife Service has only granted endangered status to 378 species and threatened status to 112 when thousands of animal species are on the verge of extinction. At the same time, it allows oil drilling, timber cutting, hunting, and fishing on many of its more than four hundred refuge areas. The U.S. Forest Service, which manages more than 190 million acres, has become, according to the *Earth First!* journal, mainly a welfare agency for the timber industry or a rancher support group as it develops one or two million acres of land a year. Finally, the U.S. Park Service has turned into "an ally of a moneyed special interest: entrepreneurs of industrial tourism" in the major national parks, building roads into wild areas and leasing facilities to private concessionaires at fire sale prices.[68] These federal agencies, in turn, have their bureaucratic equivalents in many state governments, which practice similarly destructive policies on state-owned lands, forests, and wetlands. Rather than protecting wilderness areas, all of these state agencies are seen as merely managing the decline and disappearance of Nature in pursuit of "the multiple use" doctrines of public lands management.

It is this alliance of bureaucratic power with corporate capital in global markets that Earth First! resists with its direct action and educational campaigns. Even though they often are internally inconsistent, these general points of protest sum up Earth First!'s philosophical program. There is a critique of multinational capitalism and the modern bureaucratic state in Foreman's thinking, but it mainly restates commonsense working-class complaints about everyday life rather than rearticulating any elaborate theoretical design from Marx or Marcuse. Given that its founders developed in the same era as the antiwar, civil-rights, and feminist movements, Earth First! is surprisingly free of almost any 1960s New Left rhetoric in its discourse on society, politics, and the economy.[69] In fact, Foreman ties Earth First!'s political activism back to the American—not the Russian—Revolution. To him, all of Earth First!'s activities are "the type of thing that started our country," because in his eyes, the Boston Tea Party was "probably the classic art of monkey wrenching."[70] Foreman does not mean that the Sons of Liberty were fighting to prevent tea bushes from being plucked of their leaves, but rather that the Boston Tea Party was an example of popular direct action. Ultimately, it is Mark Trail rather than Karl Marx who really represents the basic philosophical spirit of most Earth First! activists. Earth First! does have a deeply rooted faith in discursive diversity and democracy. Although Foreman often saw it as "woo-woo stuff," Earth First! also has been open to pagan nature worshiping, Gaia cultists, ecofeminism, and animal rights activists, which have their own nonrational or posthumanist ways for respecting nonhuman otherness, in its organizing a popular resistance to wilderness destruction.[71]

Monkey Wrenching: The Medium Is the Message

Since the formation of Earth First! Oregon, Washington, and northern California have been the major centers of ecotage, as activists mainly directed their monkey wrenching on timber industry targets. Such actions spread into Arizona, Idaho, Wyoming, Montana, Alaska, and New Mexico. The entire number of all monkey wrenching incidents to date is not accurately known. Some Forest Service regional offices report as many as one a month, but timber industry authorities believe one of two monkey wrenching incidents go unreported.[72] Once all of the investigation, monitoring, security, and mitigation costs are tallied, each monkey-wrenching incident probably costs $100,000 for the larger society and target, while the average monkey wrencher rarely spends more than $100 per strike. Thus, one estimate puts the total costs of ecotage in the United States at $20 or $25 million.[73] This impact is felt by industry and government in other destructive projects that are not being launched for lack of funds. Ultimately, as one practitioner claims, "the ability of monkey wrenching to intimidate and unnerve the bureaucratic and the plutocratic mind cannot be measured but likely is significant. . . . Unquestionably it has over the past 10 years revolutionized the way public lands policy is made in this country."[74] This effect on the material reproduction of a five trillion-dollar economy is essentially insignificant. Yet, in the realm of rhetoric, this tiny blip on the margin of mainly small producers in the extractive industries has been transformed by media attention into a new subversive threat from within.

In most instances, it is not the actions of economic sabotage that count, but rather the discursive responses to the acts that are critical. This edge in Earth First!'s media strategy comes out in Foreman's *60 Minutes* interview. When asked if he thought that he and Earth First! seemed "dangerous," Foreman answered, "I hope that I am dangerous to apathy. I hope that I am dangerous to not taking responsibility for life on this planet. I hope that I am dangerous to the attitude that the entire Earth is just a smorgasbord table for human beings."[75] Pushed to its ultimate conclusion, the most effective defense of wilderness and biodiversity probably *would* involve killing people by the millions. As it has emerged in the United States, however, Earth First! has not endorsed this ruthless pursuit of its ultimate objectives. Instead, it has adhered to a more moderate line, stressing the activation of liberal political initiatives to implement its basic goals. Like civil disobedience in the civil-rights movements of the 1960s, monkey wrenching is meant finally to mobilize the larger public rather than to completely paralyze the machinery of mass production. Real sabotage would work on the scale of Abbey's fictional Hayduke—blowing up real dams and destroying the real technical infrastructure of human life. These actions, however, would kill and injure thousands of people. Sticking to their pledge of non-violence against people, then, almost all monkey wrenching is staged as "a propaganda of the deed." It has worked relatively well, although elements of backlash often are built into the coverage.[76] The April 23, 1990, "Earth Day Issue" of *Time*, for example, sensationalized Earth First!

even as it criticized and trivialized its tactics by claiming ordinary people "really" do more "ecodefense" than monkey wrenchers. According to *Time*'s analysis,

> Eco-guerrilla groups [such as Earth First!] have grabbed headlines by pouring sand in the fuel tanks of logging machinery and destroying oil-exploration gear. But it is law-abiding citizens, stung by the threat to their livelihood, their recreation or their family's health, who are giving the nation's environmental movement its daily, stubborn edge. . . . Although self-preservation motivates these backyard ecologists, they often take their cause well beyond the backyard, in many cases joining the loose environmental coalitions that have become major lobbying forces in state capitals.[77]

Earth First! frequently is characterized as a band of "ecoguerrillas" in the mass media, which present it as small cells of radical extremists working against the inevitable advance of beneficial progress with terrorist methods. Monkey wrenching in these interpretations is cast as terrorism or an antiprogressive form of protest, as recent media attempts to connect Ted Kaczynski, or the alleged Unabomber, to Earth First! illustrate.[78] Spiking trees is often compared to hiding razor blades in Halloween candy, closing all wild lands to any new development is derided as un-American, and destroying heavy equipment, power pylons, or high-tension electric lines is likened to PLO raids on civilians. Monkey wrenching is intended to be nothing more than a deterrent to development. As Foreman asserts, much of this is a matter of discursive interpretation: "Terror or terrorism is the chainsaw ripping into a 500 year old Douglas fir! It is the exploding harpoon going into a sperm whale!"[79] Spiking trees or harassing whaling ships provides deterrence; the threat of it is meant to keep the trees from being cut and the whales from being killed—not to hurt or injure loggers or whalers. By selectively making strategically placed first strikes and carefully warning others about virtually undetectable booby traps, ecoteurs intend to stop development. Committing mayhem and joining in vandalism are not the ends motivating ecotage. And, by delivering these local blows against local targets using local resources, they are advancing the global agendas of preserving the environment.

Earth First! does openly engage in civil disobedience, ranging from the use of sit-ins, public marches, tree sitting, and dam site occupations to seizing buildings, accepting mass arrests, and staging street theater. These techniques have become more common in the 1990s with Earth First!'s more overtly political engagement with *defending* wilderness. Along with these techniques, however, monkey wrenching, as it is defined and defended in the *Earth First!* journal discourses, becomes a new form of civil disobedience. The central tenet "is gaining public acceptance or understanding of the need to break unjust laws. If our actions are untargeted and ethically ambiguous, then we appear as hooligans and common criminals to the public."[80] Instead of attacking people or putting individuals in harm's way, Earth First!'s program of monkey wrenching always has been focused on attacking property and/or preventing the commodification of Nature by extractive industries like mining, timbering, hunting, grazing, and fishing. Therefore, the bottom line in monkey wrenching is keeping a correct political consciousness by avoiding an "us" versus "them" mentality in ecological civil disobedience. But, it also does not stress the public witnessing and submission to punishment that classic strategies of civil disobedience have practiced. "For practitioners of civil disobedience in defense of national diversity," Foreman concludes, "the fundamental issues are wilderness and wildlife. Our opponents are federal land managing agencies and resource extraction industries. . . . Accept that we are all, to varying degrees, guilty of the destruction of the Earth. Try to separate the doer from the deed."[81] Ultimately, then, the whole design of monkey wrenching perhaps can be traced back to symbolic battles over environmentalism's image in the media.

Much of the monkey wrenching mystique or redneck Neanderthal image of Earth First! stems, in part, from Abbey's and Foreman's countermoves against prevailing rhetorics of environmental resistance in the 1970s and 1980s. Eager to discredit the environmentalists' assaults on their polluting practices, large corporations frequently employed conservative lobbyists to use high-powered public relations techniques to fashion an image in the mass media of the typical "pesty" environmentalist. A derivative of the limousine liberal stereotype scathingly denounced by Wallace and Agnew in the 1960s, the environmentalist allegedly was a white, wine-and-Brie, upper-middle-class professional (who probably once was a hippie or antiwar activist) with no sympathy for the plight of the ordinary working man put out of work by the environmentalist's meddlesome "tree hugging" love of Nature (which meant clean water, clean air, and clean beaches around upper-middle-class enclaves of wealth). Played up in the Mobil fables, James Watt's lobbying, and television sitcoms, this image of "the environmentalist" plainly was used to contain real ecological concerns through the coercive cultural coding of environmentally minded individuals. It survives today in almost any focus group's association of threat, for example, with the label "Sierra Club," "backpacker," or "Greenpeace."

Recognizing how corporations often contained environmental concern in these kinds of cultural stereotypes and media myths, Abbey's Monkey Wrench Gang put a more "working-class," or even "redneck," spin on environmentalism.[82] As "authentic" working-class drifters and handymen, these images have rearranged the cultural codings of environmentalists' media images. Initially billing himself as a member of "rednecks for wilderness," Foreman continued in Earth First! much of his "buckaroo" persona from his days as a lobbyist with the Wilderness Society to counteract the image of environmentalists as upper-middle-class snobs. As one of Foreman's associates explains this element of cultural theater, "we prided ourselves on being buckaroos. We could out stomp and out drink others. We wanted to show that to be for wilderness you didn't have to be an effete Brie-eater. You could drink beer and play country music and be an outlaw. We were not giving any ground to the other side in redneck ethic. We weren't going to concede who were the cowboys."[83] Although this symbolic shift has perhaps won Earth First! many more supporters among working-class people, the image itself already set the stage for a media backlash as well as feminist protests within the movement. These supposedly more positive images of buckaroo civil disobedience can also be wrapped back around negatively as signs of an "ecobrute" or "caveman," which ties into many other dangerously defined working-class stereotypes, such as "cowboy," "redneck," "lowlife," "populist," or "hooligan." In the current media environment, this rhetorical counterattack has not always succeeded in all media markets, but the symbolic basis for its success among many target groups exists. As Bowden aptly observes,

> in the marketplace of ideas their image [Earth First! monkey wrenchers] functions as a kind of pornography for the largely well-educated, well-heeled middle and upper-middle class American environmental movement. The Fantasy of Direct Action Against The Beast, Babylon, The Military-Industrial Complex. . . . We pretend to be puzzled why we hesitate at doing similar things. But we are not puzzled. We are fascinated by monkey wrenching because we feel powerless. We have made an issue out of a fantasy because we crave that one moment of smashing a blow.[84]

Angry individuals, fearful about their ecological future, enjoy day-dreaming about delivering a crushing strike against the technological empires that now are ruining everyone's environment. Likewise, fearful corporations, angry about being cast as producers of waste and poison, benefit from recounting nightmares of crazed radicals destroying the lifelines that sustain the average American consumer's "high standard of living." In this rhetorical impasse, the existing corporate system, the state apparatus, the viewing/consuming public, and Earth First! all directly enjoy many satisfactions from symbolic representations of monkey wrenchers as "ecoguerrillas" or "ecoterrorists."

With regard to their media strategies, then, Earth First!ers clearly recognize that

> the radical environmental message, whether concerning old-growth or dolphins, would not be receiving the widespread coverage it is today were it not for the "publicity value" of monkey wrenching. Most of the coverage EF! has received over the years—in *Esquire, The Amicus Journal, The Nation,* local newspapers, etc.—has concentrated on ecotage, often favorably or at least without overt condemnation.[85]

Before his arrest in May 1989, Foreman rarely made it into local or regional, much less national, news broadcasts. After his incarceration, *60 Minutes, Rolling Stone, Smart, Spin, Buzzworm, Time, Newsweek, Smithsonian, E Magazine,* CNN, and the major networks all gave Earth First! and/or Foreman plenty of airtime and page space.

By staging events that both draw major media coverage and popularize Earth First! positions for large national and international audiences, Earth First!ers try to influence public opinion directly to speak against the incursions of the state and corporate experts into the everyday life world. Media attention often stigmatizes Earth First! ecotage in terms of needless property destruction or dangerous terrorism instead of civil disobedience or ethically responsible acts of conscience. Ironically, this coding of Earth First!'s monkey wrenching has not completely contained this symbolic threat to the established order. Instead, such interpretative frames seem only to have won Earth First! even more popular sympathy and to have assured a continuing media watch on its activities. Media attention has had its costs, but the benefits have been much greater. The media's fascination with monkey wrenching

> takes seemingly obscure environmental issues out of the dark of scientific calculations into the limelight of individual passion and commitment. Even when the media distorts it by emphasizing its unlawfulness rather than its motives, it has been, like civil disobedience, an important element in the broader campaign to rally public opposition to wilderness destruction. For instance, in news coverage of the old-growth controversy, the subject of ecotage inevitably comes up, along with civil disobedience, and it's clear this activism attracted the coverage in the first place.[86]

Most monkey wrenching-minded Earth First! activists realize that "Earth First! is guerrilla theater, not guerrilla war, and it can be a dull, hard life."[87] In the real world, Earth First!ers live almost hand to mouth on the edge of real poverty, struggling to protect small parts of their everyday life world from commodification and rationalization by big business and the state. The experience of a local organizer in New York sums up these more cautiously implemented political goals and ecological values:

> EF! is more than pickets and demonstrations. It's an attitude—no more compromise. And there are so many ways to get this message across. There are definitely times and places for demonstrations. . . . But even more important is becoming informed and articulate on issues, then countering development with no-compromise, deep ecology messages

at every juncture. And we can do this constantly, through letters to editors, position statements, talks to meetings, wherever there's an opportunity to speak out. So even if you're not a spike-wielding demonstrator, this movement has a crucial place for you. . . . Learn about your local issues. Then, get the group members to start thinking about these issues from the deep ecology perspective. Prepare proposals from a no compromise position. If you go to meetings, speak or testify as an EF!er. Explain that wild areas and wild organisms need to be saved just because they are. . . . Finally, think awful hard about monkey wrenching. Let's use this tactic the way it was always meant to be used—as a statement when there's no other way to make the statement, or as a last resort.[88]

Here the localistic, particularistic concern for biocentrism and biodiversity, expressed in more thoughtful but still noncompromising public actions, calls for serious consideration of monkey wrenching as a strategy. It is meant to be a personal statement of the last resort put within a larger democratic rhetoric of morally aware civil disobedience. As one Earth First! activist sums up the final motive behind these politically driven media campaigns, they grab people's attention: "We want people to have the jitters. They ought to have the jitters. The planet is being killed by corporations right now."[89]

Summary

Earth First! still stands for protecting the Earth and all of its life-forms from further economic growth. Yet, this global agenda takes open-ended, localistic forms of implementation. In the group's discourse, "Earth" is a region, a site, a seascape or landscape known to a definite group of people rather than a dead astronomical abstraction of the entire planet. It is something manageable and immediate that takes an identity of the earth and makes its apparent good or interests first in the objectives of the group. "Earth" in this form also is postnational, post-statist, postbureaucratic in its social conception and cultural understanding. Its defense, once an activist or local group adopts such discursive understandings of "Earth" and its endangered state, can be endlessly improvised on a local, decentralized, individual basis, making its national and transnational popularization extremely easy.

Earth First! tactics probably cannot spread everywhere. This movement presumes a minimum level of personal wealth and leisure time
to engage in civil disobedience and/or monkey wrenching that is not
commonly enjoyed in many regions around the world. The concern for
Nature and general biodiversity also are not pressing concerns in many
areas where human survival is a daily preoccupation. Earth First!'s philosophy tends to accept human extinction in such ecologically stressed
regions to give animals, plants, and local ecosystems a chance to recover.
An openly repressive government or business community, which would
ordinarily accept using deadly force against ecological activists, also
could prevent Earth First! groups from acting out their unusual forms
of activism as "civic politics" in many countries around the world.[90]
Hence, it is not too surprising that outside of certain areas in the United
States, Earth First!'s efforts to defend Nature and its members' perceived
place within natural ecosystems have taken hold mainly in liberal Western democracies like Canada, Australia, Sweden, Germany, and Great
Britain.

Earth First! does have a fairly telling critique of today's industrial
society. The larger design behind advanced capitalist industrialism has
been directed at attaining greater technical efficiencies in its instrumental rationality to provide certain "public goods," including more social
welfare services, political stability, economic growth, and mass consumption. This instrumental rationality of advanced capitalist industrialism, then, as it has grown on a transnational scale, has destructively in-
vaded more areas of everyday life and fused political administration with
economic management, actually undercutting the attainment of its allegedly original substantive ends of emancipation and prosperity. It has
confounded its own means/ends efficiencies.[91] The transaction costs of
instrumentally rationalizing everyday life on a global scale increasingly
decrease the benefits of making the transactions, which imperils the entire life world of Nature and society. And in turn, more and more "public bads"—pollution, rising prices, resource scarcities, national conflicts,
great-power competition, domestic violence, unemployment, or service
shortages—are created in this process of providing fewer and fewer
"public goods."

Earth First! understands that these rationalization crises entail a final invasion of the last refuges of Nature to sustain the prevailing regime of high-tech industrial productivism. The existing ways that industrial output is produced lead to less biological diversity, and the established definitions of what constitutes output accept turning wilderness into "natural resources." What is produced is more needless waste, fewer bio-diverse ecosystems, and more artificial environments. And, finally, who decides on these development policies are not the people, animals, and plants who will be destroyed, but small cliques of self-interested corporate managers and insensitive government bureaucrats, who legitimate their actions with technical discourses about more productivist growth.[92] In the long run, Earth First!'s challenge to the cultural logic of modernity might provide a truly radical alternative to the policies of The Nature Conservancy discussed in the next chapter. Instead, Earth First! too often plays into the hands of the technocratic corporate and governmental modernizers, as the FBI counterintelligence operations against Dave Foreman in 1989 or the local police harassment of Judi Bari and Darryl Cherney during 1990 illustrate. Thus far, the inherent irrationalities in contemporary transnational exchange, which Abbey and Foreman so clearly describe, have only sparked Earth First! to mount limited oppositional attacks, which leaves these ecoraiders operating at best as weak countervailing forces against the increasing pressures of global competition, performance, and accumulation.

3

The Nature Conservancy or the Nature Cemetery: Buying and Selling "Perpetual Care" as Environmental Resistance

Unlike Earth First!, many find it difficult to criticize the work of The Nature Conservancy (TNC). Compared to many other environmental organizations, it is doing something tangible, immediate, and significant to protect Nature—buying, holding, and guarding large swatches of undisturbed natural habitat—by sticking to the ground rules of the current capitalist economy. Millions of acres, occupying many diverse ecosystems all over North and South America, now are being held in trust by TNC. This trust is being exercised not only for future generations of people, but also for all new generations of the plants and animals, fungi and insects, algae and microorganisms inhabiting these plots of land.

The Nature Conservancy's roots can be traced back to 1917 when a study group at the University of Illinois formed within the Ecological Society of America under the name of the Committee for the Preservation of Natural Conditions.[1] Concerned about the rapid spread of ecosystem destruction in the United States, this committee conducted a long scientific and intellectual debate about how to counteract this environmental damage. Frustrated by a lack of action on the part of the committee, a dissident faction broke with the Ecological Society and established a new body, the Ecologist's Union, in 1946.[2] One of its members, Dick Pough, visited the United Kingdom and looked into the Nature Conservancy programs administered by the British Government. When he returned to the United States, Pough encouraged the Ecologist's Union to emulate the English program's policies for preserving lands as open spaces and wildlife havens. Pough lobbied the union to assume the name of The Nature Conservancy, but he wanted it to avoid the public funding formula used in England in favor of keeping the organization private and soliciting contributions from corporations, foundations, and wealthy individuals.[3] In 1951, the Ecologist's Union renamed itself The Nature Conservancy, and Pough approached Mrs. DeWitt Wallace of the *Reader's Digest* for a donation. With two hundred thousand dollars from Mrs. Wallace, The Nature Conservancy was in business. Its first purchase of sixty acres was made in 1955 in the Mianus Gorge along the Mianus River, north of New York City, which had grown in size to 555 acres by 1994.[4]

Since 1955, TNC has been putting the donated dollars of its members to a great deal of good by buying and selling tracts of land to protect them from development. Beginning with the sixty acres in the Mianus River Gorge, this organization has protected by direct acquisition and trust negotiations more than 7.5 million acres of land in North America as well as Central America, South America, and the Caribbean in more than ten thousand separate protection actions. In the past four decades, on pieces as small as a quarter of an acre to as large as hundreds of square miles, The Nature Conservancy in the United States has arranged for the ongoing protection of an area the size of Connecticut and Rhode Island.[5] Although this success is plainly not everything that needs to be done, it is something to admire inasmuch as these moves were what could be done under current economic and social conditions. Given that so many ecological initiatives frequently fail, this string of successes cannot be ignored.

One also must admit that The Nature Conservancy's achievements are seriously flawed, even if these flaws may reveal much more about the private property and free enterprise economic systems it must work within than they do about the organization itself. Consequently, this chapter takes the actions of The Nature Conservancy as a marker of other major structural shifts within the political economy of advanced capitalist societies. Because of what has happened to Nature, how capital operates, and where resources for change must be solicited, TNC does what it can. It pragmatically, or perhaps uncritically, accepts these realities as background conditions, and then tries to do something positive within the constraints imposed by these limitations. Yet, as a result, the tenets and tenor of the Conservancy's operations as "an environmentalist organization" are those of almost complete compliance, and not those of radical resistance to this system of political economy. In the final analysis, one must rethink how each real material advance of TNC toward protecting land as real estate might actually be a serious reversal for those supporting the cause of Nature preservation as one that should never accept reducing Nature to real estate.

The Nature Conservancy as REIT

The Nature Conservancy as a not-for-profit civic organization works the food chains constructed in contemporary society by the direct mail technologies of psychodemographic profiles, address lists, and lifestyle marketing. Like other similar predators in the political party, interest group, life insurance, book club, upscale catalog, and mutual fund families, TNC baits its prey through the mail. Indeed, the strategies of recruitment mobilized by TNC are not unlike those deployed by scores of life insurance or mutual fund operations, targeting the same demographic target groups with their direct mail campaigns. As you open the envelope, a big bug-eyed bird, or a native American sandhill crane, already is ogling you on the envelope and the papers comprising the prospectus, which are soliciting only ten dollars to protect this bird's nest egg from farming, cities, and subdivisions causing wetland destruction. The potential member is addressed sincerely, as in a mutual fund package, as "Dear Investor," and invited to invest cold hard cash in "prime real estate, if you're a crane or a bass or a sweet pepperbush or a redwood. Or a toad or a turtle. And a lot of it's nice for people too— lovely deserts, mountainsides, prairies, islands."[6] The message is plain: this group obviously is not just into monkey wrenching or funny business, like "the more visible and vocal conservation groups."[7] Instead, this organization offers its members, in a sense, individual participation in a unique kind of green real-estate investment trust (REIT), one with a nicely diversified portfolio in scenery, good vegetation, habitat, spawning grounds, and water, because this mechanism is the "unique, inexpensive, and *effective* way" the Conservancy goes about its "nonprofit business" of conserving Nature by letting "money do our talking."[8]

With fifty-nine chapters in fifty states, TNC boasts, "check your phone book." This shop is not a fly-by-night boiler room scam. It will be around and always ready to serve any of its members with tours, glossy magazines, and chapter newsletters, all detailing the health of this unusual, but still green, REIT, because "bargains in diverse real estate are what we look for and find."[9] The arresting image of the sandhill crane in distress will remain with you after you write the check, because as your personal reward for joining this movement you receive some lovely welcoming gifts from The Nature Conservancy: first, a "complimentary sandhill crane bookmark" so that you will "be reminded of your personal commitment to the preservation cause," and, second, a "sturdy National Conservancy tote bag featuring that bug-eyed bird— the American sandhill crane. What a great way to conserve resources and avoid using plastic!"[10]

With such marketing savvy at its command, why does this group bother with "small donors" or "little people"? It is simple: TNC admits that it must create its bona fides from large numbers:

> We need you, very much! Those hardheaded foundations, corporations, and individuals who give us money or property must be convinced that our ranks include a lot of intelligent, concerned, articulate citizens: people who know that the natural world and all it harbors needs our help. *Yes*, we need you and your ear and your voice—*and* your $10 ($10 times 700,000 members helps protect a lot of acres).[11]

Once you join, you can sleep more securely, knowing that you and seven hundred thousand other fellow investors in Nature's nest eggs are letting your money do your ecotalking for you. That river of cash is vital. It is the eternal spring needed to replenish the Conservancy's revolving fund, "every dime of which is plowed in the unpaved and as yet unplowed."[12]

Keeping the earth green means keeping the green flying into The Nature Conservancy's coffers, so the bottom line is simple. The direct mail package suggests: "Enclose a check for $10 in the return envelope. (Send more, if you can spare it.) NOTE that it's tax-deductible [except for the $6.95 retail value of the TNC tote bag]. Mail the form. Go."[13] So, now, like a piece of the rock or the peace of mind that quality buys, the average consumer becomes one with that elegant TNC oak leaf in green or black festooning every shard of Nature Conservancy printed communication. As an enterprise in pursuit of peace of mind, TNC is "dead serious about hanging on to nature's balance," but "not by campaigning or picketing or suing."[14] These brash tactics get you on the evening news, but they do not protect habitats from development. To beat this system, TNC whispers to you in an honest aside, one just about has to join it. So, to save the sandhill crane, the Conservancy will "*just BUY the nesting ground*"[15] that the crane needs for its nest egg, guaranteeing that another natural tenant gets needed floor space in the big real estate range of Nature to justify the conservancy mission of this unusual mutual fund. As you return your application and ten-dollar check to this green REIT, you learn that the sandhill crane was a temporary check-in at the International Crane Foundation of Baraboo, Wisconsin. Now, as the Conservancy's direct mail letter signs off, we are left thinking: "Maybe he's smiling. Certainly he will be if, after he leaves the Foundation, his first motel stop has been reserved with your $10."[16] So, as a Conservancy investor, one now acquires access to all of those special lands mutually funded by seven hundred thousand other like-minded investors, as well as a contractual tie to swell tenants like this American sandhill crane.

Capitalist Anticapitalism

TNC is quite interesting as an environmental defense organization in that it turns the reigning strategies of land commercialization under capitalism for profit into a new set of legal tactics for land de-commercialization under environmentalism in not-for-profit "protection actions." These actions, however, are not frenzies of reckless expenditure by TNC, because it also prides itself in being able to say "we don't just shovel cash at the problems."[17] Instead, the Conservancy tends to serve as a broker, taking lands out of circulation in the market and entrusting them to third parties, such as "states, universities, other conservation groups—any responsible organization that can care for and protect the land from anyone."[18] The Conservancy's bag of tricks must exploit the full range of legal options available in the real estate industry: "we buy some lands, trade for others, get leases and easements, ask to be mentioned in wills."[19]

In The Nature Conservancy's operational codes, therefore, a triage system immediately comes into play, following from its flow of donations. Some lands of Nature are more "ecologically significant," some regions are much more "natural areas," but some grounds are far less "protectable" than others. Once again, the methods of the Conservancy show how it implicitly sees Nature as real estate properties inasmuch as its chapters must constantly triage the acreages they receive—labeling some as truly ecologically significant, some as plainly natural areas, some as merely "trade lands." The latter are denaturalized zones or artificialized spaces whose main value is that they can be sold, like old horses for glue or worn-out cattle for dog food, and the proceeds used elsewhere to promote conservation. Strangely, in seeking to preserve Nature, TNC oversees its final transformation into pure real estate, allowing even hitherto unsalable or undeveloped lands to become transubstantiated into "natural areas" to greenbelt human settlements and recharge their scenic visits with ecological significance.

TNC obviously is an operation pegged at winning supporters and recruiting members among the owning classes or, at least, the would-be and soon-to-be owning classes. To support the organization, the executive directors do not want its members to push lawsuits, organize protests, or picket supermarkets. Instead, they appeal to professional-managerial class consciousness, encouraging new members through helpful financial hints in Conservancy publications to support The Nature Conservancy by giving it land, assigning it common stock shares, remembering it in their wills, writing checks to augment the group's cash flow. This is an organization whose actions are no-nonsense. Like any other professional, it cuts to the chase, soliciting money from its backers, and not more messy contributions—time, labor, or moral witness.

As it adopts these tactics, TNC must recruit such unlikely allies as trust attorneys and tax accountants to conduct its resistance on behalf of Nature. With more than 50 percent of its protected lands coming through donation, the Conservancy must enhance the green-leaning property owner's propensities to do good by making it cash-out on the donor's bottom line. On the one hand, the Conservancy assures its marks that "making a gift of land to be preserved in its natural state brings the satisfaction of protecting wildlife and creating a sanctuary that future generations will cherish," while, on the other hand, observing that gifts of "real estate can carry substantial tax advantages" inasmuch as such gifts permit ecologically minded donors to avoid "the capital gains tax and receive a charitable deduction for the full fair market value of the property."[20]

The Conservancy's tax experts, of course, do not restrict their advice to land alone. As long as individuals continue to be so important to the Conservancy's cash flow, providing more than 70 percent of all funds, members are invited to include the Conservancy in their wills, to assign heavily appreciated stocks to the Conservancy's war chest, and to give ever greater amounts of cash to the Conservancy every year. Remembering the Conservancy, then, also becomes a sign of shrewd estate planning since "bequests to the Conservancy are entirely free from federal estate tax and can therefore offer substantial estate tax savings."[21] Similarly, signing over that steeply appreciated stock, once converted to a Nature Conservancy gift, "can be particularly tax-wise" by avoiding capital gains tax plus getting a charitable contribution reduction. And, any extra cash burning a hole in the donor's pocket can go to state chapter coffers, allowing one to ascend from ordinary membership into the ranks of the Acorn Club ($100 or more a year), the Mallow League ($100–$149), the Songbird Society ($150–$249), the Coneflower Club ($250–$499), the Bobcat Club ($500–$999), the Bittern Club ($1,000–$2,499), the Orchid Society ($2,500–$4,999), the Fallon Club ($5,000–$9,999), or the pinnacle of Conservancy generosity, the Eagle Club ($10,000 or more).

All of these different ways of giving are arranged so as to be flexibly targeted. The development directors of the Conservancy want to maximize freedom of choice for donors, who worry about how and where their cash or assets might protect natural areas. So, the donor, "as with all donations to the Conservancy," is given considerable discretion to designate "where and how you wish your gift to be used."[22] The triage policies of the Conservancy, however, will limit discretion inasmuch as the lands given to the organization may not have any "ecological significance" for its purposes. It can, and does, sell off lands as "trade lands" that the original donors might have seen as ecologically significant. Often these sales ruffle some feathers, but in the end the Conservancy justifies it as promoting its goals of protecting biodiversity and providing sanctuary for rare ecosystems, even if some donor might have wanted his or her heirs to enjoy that ground in an undisturbed state. It is on this ground, then, that the Conservancy really wins many of its battles in skirmishes over estate planning, tax avoidance, and charitable giving. Instead of picketing and suing, its personnel are directed into "development" to find the real estate and cash needed to wage the Conservancy's battles for biodiversity.

Conservancy-wide private donations still come overwhelmingly from individuals. Despite all their talk about doing everything they can to advance environmental protection, corporations only provided 11.5 percent of The Nature Conservancy's privately donated funds in 1993. Various foundations contributed 15.5 percent, but 73 percent of all funds from private sources still come from individual donations.[23] These facts underscore the shallowness of corporate commitment to the Conservancy's goals, while at the same time illustrating the centrality of identifying fresh targets, recruiting new members, and soliciting cold cash in the direct mail wars in The Nature Conservancy's operations.

Corporate gifts also rarely amount to much more than another maneuver in public relations campaigns to prove how, as "corporate citizens," big companies seriously take their obligations to be responsible "stewards of the environment." Thus, for example, a multibillion-dollar operation like Norfolk Southern Corporation can commit one hundred thousand dollars in 1994 over five years "to fund acquisition and stewardship of natural areas" all across Virginia.[24] This contribution followed a smaller 1990–91 gift of fifty thousand dollars to help the Conservancy acquire 1,700 acres in Bottom Creek Gorge near Roanoke, Virginia, where the railroad still has a major operations base. All of this must be applauded, but it is a small victory. It comes after many big defeats when one puts it in the context of Norfolk Southern's continuing operation of hundreds of coal trains pulling millions of tons of coal out of Virginia mines to be burned in mostly inefficient boilers all around the world to make large quantities of electricity and pollution. Thousands of acres of land, scores of watercourses, and dozens of railway yards are all being continuously polluted and degraded every day along hundreds of miles of track as the direct result of this corporation's railway business. In the frameworks of TNC, these properties are not ecologically significant because they are lands already occupied by trade. The Conservancy basically ignores the big dollar problems developing there, and instead solicits nickels and dimes to buy pristine hollows still out of harm's way. Not surprisingly, Norfolk Southern and other corporate citizens always are "pleased to once again support the work of the Nature Conservancy,"[25] because in many ways donations to these special showcase acquisitions, such as Bottom Creek Gorge, are just chump change for making chump changes in big businesses' exploitation of the environment. Corporate bosses can justify the donations to themselves and their stockholders as an "effort to protect our valuable natural resources," and The Nature Conservancy deposits these generous gifts to pay for preserving "the most important concentrations of rare species and natural communities" in their highly selective zones of conservancy.[26]

To appeal successfully to corporations and foundations, however, the Conservancy leaves its visions of the threat to "Nature" very vague. When it asks for land to protect wildlife and create sanctuary for ecosystems, it tends not to detail the ultimate cause of its concern. Protect it from what? Create sanctuary from what? The answer is the same economy that is allowing its members to accumulate stock, mail in cash donations, buy and sell land as private property. In many ways, the Conservancy is disingenuous in its designation of only some of its lands as "trade lands." Actually, all of its protected lands are trade lands, trading sanctuary and protection here (where it is commercially possible or aesthetically imperative) to forsake sanctuary and protection there (where it is commercially unviable or aesthetically dispensable). It extracts a title for partial permanence from a constant turnover of economic destruction anchored in total impermanence. Thus, the Conservancy ironically fights a perpetually losing battle, protecting rare species from what makes them rare and building sanctuary from what devastates everything on the land elsewhere with the proceeds of its members' successful capitalist despoliation.

TNC necessarily embraces the key counterintuitive quality of all markets, namely, a dynamic in which the pursuit of private vices can advance public virtues. It agrees to sacrifice almost all land in general to development, because it knows that all land will not, in fact, be developed. On the one hand, excessive environmental regulations aimed at resisting the development of all land might destroy this delicate balance in land-use patterns. In accepting the universal premise of development, on the other hand, it constantly can undercut economic development's specific enactments at sites where it is no longer, or not yet, profitable. Some land will be saved and can be saved, in fact, by allowing, in principle, all land to be liable to development. It needs trade lands to do land trades to isolate some land from any more trading. In allowing all to pursue their individual vices and desires in the market, one permits a differently motivated actor, such as The Nature Conservancy, to trade for land, like any other speculator, and develop it to suit its selfish individual taste, which is in this case is "unselfish nondevelopment." This perversely antimarket outcome satisfies the Conservancy's desires and ends, while perhaps also advancing the collective good through market mechanisms.

Acknowledging the Conservancy's public virtues must not blind one to its facile dealings in private vices to advance its cause. The Nature Conservancy "succeeds" in its own way because it accepts so much of what so many other ecological groups ardently picket, sue, and protest to stop—the market economy and corporate capitalism. By bending corporate techniques for profit to nonprofit ends, it suspends a few of the marketplace's most destructive outcomes in a few far-flung corners of the environment. But, to attain this success, it also is always all business, even if it turns its businesslike bearing against excessive business depredations. For its protection benefits to hold fast somewhere, it must accept the dictum that the marketplace will rule everywhere. The market rules even in Conservancy lands, because these plots are mostly marked out in market-mediated trades.

Death of Nature/Birth of Biodiversity

Until the early 1970s, TNC adhered to many essentially romantic assumptions about Nature in its various operations. Nature existed as a real presence in places left undisturbed by human beings.[27] At such sites, therefore, all plants and animals embodied Nature's substance, and as such they were all worthy of protection. Therefore, acquiring any and all land that was not now being touched by some human settlement's economic and social activity constituted a meaningful act of Nature conservation. The battle lines were drawn across territories understood as land under active current social utilization (or artificial zones without a large Nature presence) and land no longer or not yet put to such use (or natural zones without a large social presence), making it, therefore, worthy as a site of Nature conservancy.

At some point in the 1960s and 1970s, these discursive frames became obsolete, even though many people today inside and outside of the environmental movement do not recognize their obsolescence. Bill McKibben and Carolyn Merchant, to a very real extent, are right.[28] Nature has ended. Nature is dead. Perhaps it was the conquest of extraterrestrial spaces in the moon landings of 1969–70, whose televisual and photographic images of the Earth sublated conceptual antinomies like Nature/Society, out-there/in-here, wilderness/settlement, and environment/economy in pictures of one undifferentiated little blue planet floating in space. Perhaps it was the penetration of transnational capitalism even into Moscow and Beijing during Nixon's détente, whose acceptance of Pepsi, Kentucky Fried Chicken, or Pan Am passengers underscored how commercial commodities now could go anywhere anytime anyone wished them to do so. Perhaps it was the collective shudder of the 1971–73 *oil shokku* from OPEC, whose antics illustrated how tightly coupled all human societies were to petrochemical energy supplies mostly pumped up from under a handful of Mideastern deserts.

The brittleness in the traditional categories, pitting "Nature" against "Society," finally snapped. There really were no lands without any traces of some large social presence. Hundreds of individual plants and animals were not merely being disturbed by humanity; they were being eradicated as species in wholesale extinctions. Scientific observations of air quality, water quality, atmospheric integrity, bone composition, or milk production were indicating that human beings have profoundly disturbed what had been regarded as unalterable sovereign Nature with industrial pollution, greenhouse gases, chemical contamination, and radioactive wastes. Hence, to pretend to be conserving "the untouched and undisturbed expanses of Nature" in simple actions of land ownership made very little sense. Nature, as it had been conceptualized by the Ecologist's Union in 1945 for The Nature Conservancy to preserve after

1951, had been disappearing right from under everyone's noses as global modernization accelerated at even greater rates from 1945 to 1973. Once-undisturbed lands could no longer be regarded as the container of Nature as such, they had to be continuously reassessed to determine their ecological values, creating the space of surveillance, as chapters 4 and 5 argue, known as "the environment."[29]

From 1951 to 1973, TNC did have a decidedly scientific air about its operations, even though with its haphazard policies of property acquisition the organization still worked as a land collection program to preserve scenic beauty, rare plants, and wild animals on whatever land it would get whenever it could get it. Low-key and nonconfrontational, The Nature Conservancy's style of environmental protection had been fairly cozy with big business from the beginning. Under the leadership of Pat Noonan from 1973 to 1980, however, this almost accidental precedent was turned into an article of faith. During the intensive environmental conflicts of the early 1970s, as Noonan notes, "corporations and environmentalists were butting heads, but we knew the free-enterprise system was a fantastic motivator. So the Conservancy decided to reach out to corporate America. No other environmental group was doing it."[30] This operational strategy was augmented with a new mission statement from Bob Jenkins, The Nature Conservancy's in-house biologist. Dissatisfied with the Conservancy's acceptance of land with dubious ecological worth, although it often appeared to be undisturbed as natural scenery, Jenkins saw biodiversity rather than natural appearance as that quality which TNC should be preserving.[31] Instead of protecting any and all land that it acquired, Jenkins's strategy stressed preserving only those lands populated by rare or endangered flora and fauna in unique ecosystems. Scientific surveillance was combined with high-test property management to give TNC a much clearer purpose.

Noonan's acceptance of corporate free enterprise and Jenkins's agendas for surveying, inventorying, and guarding biotic diversity recentered The Nature Conservancy's ideologies and institutions on the strategies they continue to follow today. Grove concludes:

> The Nature Conservancy would henceforth seek out and attempt to save species and biotic communities that stood in danger of disappear-

ing under human pressures. It would buy land when necessary, but to cut down on management and overhead costs, property was often turned over whenever possible to responsible federal and state agencies for protection. Further, TNC would decentralize, establishing state field offices that could keep closer watch on the land. Coinciding with its new focus, the national staff grew out of its basement quarters in Washington and moved just across the Potomac River into a high rise office building in Rosslyn, Virginia. Emerging as well in the months and years that followed were the practical arguments for maintaining biotic diversity. . . . Although morals and aesthetics still apply, the emphasis turned to "save this plant because we may need it someday" instead of "save this plant because we have no right to kill it."[32]

Over the past two decades, TNC has grown by leaps and bounds by sticking to the operating objectives in this "preserving biodiversity" mission statement.

Because Nature has ended, material signs of its now-dead substance need to be conserved as pristine preserved parts, like pressed leaves in a book, dried animal pelts in a drawer, or a loved one's mortal remains in a tomb. Nature is dead, but long shall Nature live in the environmentalized forms of rare species, exotic biodiversity, land preserves, and threatened ecosystems. As powerful anthropogenic actions have recontoured the earth to suit the basic material needs of corporate modes of production, these artificial contours now define new ecologies for all life forms caught within their "economy" and "environment." The "economy" becomes a world political economy's interior spaces defined by techno-science processes devoted to production and consumption, while "the environment," in this sense, becomes a planetary political economy's exterior spaces oriented to resource creation, scenery provision, and waste reception. The pickets and suits deployed by other more vocal ecological groups perpetuate a faulty one-dimensional consciousness, "believing that environmental protection and economic growth are somehow mutually exclusive."[33] These antiquated perceptions cling to the illusion that Nature is alive, and somehow avoiding its subjection to capital in the commodity form by remaining wilderness. Recognizing the true totality of transnational capital's power, which easily commodifies wilderness in many environmental products, the goal of "The Nature Conservancy is to change this unfortunate perception," because, as anyone attentive to capital's dynamics can attest, "our economy and environment are not antagonists, they depend on each other. . . . protecting our natural resources generates economic benefits."[34]

Natural resources exist, but Nature does not. Economic survival is imperative, but the commodity logics driving it need to be grounded in sound ecological common sense lest the limitless dynamism of commodification be permitted to submit everything to exchange logics immediately. Time is now what is both scarce and centrally important in the highly compressed time-space continua of contemporary commodity chains. It is no longer a question of jobs versus the environment, because fewer jobs will not resurrect Nature. Nature is dead, and the environment generating global production assumes that jobs are necessary to define it as the space of natural resources. Doing jobs irrationally and too rapidly, however, is what destroys these environments, making jobs done rationally and at an apt pace ecologically acceptable. Consequently, the agendas of environmental protection must center on the "question of the short-term vs. the long-term," and this is "what the Conservancy is all about."[35]

At its best, The Nature Conservancy's ideology is just plain old common sense. It does not take a genius to connect causally the presence of dead fish downriver to the location of some chemical plant dumping industrial effluent upriver. It may be true that many Masters of Business Administration have not been able to draw the lines of causation between such environmental dots, but in all fairness to them, their being in the business of administering capital's mastery of the economy and society all too often recused them of this responsibility. The Nature Conservancy's directors, then, wish to serve as outside consultants to correct this oversight as they settle for pushing green common sense, or working with business communities "to promote the fact that a healthy economy and a healthy environment go hand in hand."[36]

Therefore, TNC is never fanatical about the natural purity of the lands it chooses to protect. If it cannot have untouched lands, it will accept highly touched territories to advance its environmental goals. In Virginia along the James River, for example, the Conservancy during 1994 brokered a three-way deal between itself, the James River Corporation, and the American Farmland Trust on the Upper Brandon Plantation, which has been under cultivation since 1616. The accord provided a conservation easement with significant tax benefits for the James River Corporation in exchange for the perpetual maintenance of the plantation's 1,800 acres as open space, wildlife habitat, and environmentally sound farmland.[37] With 581 acres of tidal marshes and bald cypress swamp, the plantation is a vital area for maintaining many fish, fowl, and mammal populations in the James River watershed and Chesapeake Bay ecosystem. With the tax and public relations payoffs for the James River Corporation, which agreed to forego extracting $8 million of sand and gravel deposits on this property, and these wetlands protection payoffs, this deal is being highly touted by The Nature Conservancy as one "benefiting the balance of nature and the balance sheet."[38] In reviving "the profitability of the farm with environmentally sound, yet economically viable agricultural practices," the James River Corporation and The Nature Conservancy are celebrating this arrangement on the Upper Brandon Plantation as "a model for land stewardship throughout Virginia and the nation."[39] Raw Nature has been gone along the James River for nearly four centuries, but the highly processed environments that human economies and societies have substituted in its place still need to be conserved according to a proper green code of conduct. Hence, the Conservancy can intervene in the name of Nature, which now is marked on the Upper Brandon Plantation at best by waterfowl passing through on their still uncommodified flyways, to reconcile these double-entry books for business and the environment.

Beyond Perpetual Care

The Nature Conservancy incessantly touts itself as a not-for-profit provider of "perpetual maintenance" on its many plots of wild land, promising to maintain their natural qualities in perpetuity. This mission statement may bring peace of mind to its members, but it sounds oddly like the sales pitch used by burial societies or memorial parks. On the one hand, it aptly frames one of the organization's most vital manifest roles, namely, safeguarding peace and quiet for birds and bees in undisturbed preserves. Yet, on the other hand, this orientation perhaps reveals the organization's more significant latent role, namely, serving as "the Nature cemetery" rather than "the Nature conservancy."

Somewhat bizarrely, this outcome squirms around inside the organization's many operations. Nature, in all of its wild mystery and awesome totality, is not being preserved. It is, in fact, dead, as McKibben and Merchant tell us. Nonetheless, its memory might be kept alive at numerous burial parks all over the nation where glimpses of its spirit should be remembered by human beings in a whiff of wildlife, the scent of a stream, or the aroma of surf. This goal may be a well-intentioned one; but, in many ways, all that The Nature Conservancy does boils down to serving as a burial society dedicated to giving perpetual maintenance and loving care at a variety of Nature cemeteries: Forest Glen, Mountain Meadow, Virgin River, Jade Jungle, Prairie View, Harmony Bay, Sunny Savannah, Brilliant Beach, Desert Vista, Happy Hollow, Crystal Spring. As Nature's death is acknowledged, more and more plots are needed to bury its body in gardens of eternal life. Thus, the call for members, funds, and donations will always grow.

This mission is even more ironic given the means whereby it is funded. Those humans, whose production and consumption have had so much to do with Nature's death, the middle and upper-middle classes, are given an opportunity to purchase some atonement for their anonymous sins as consumers by joining The Nature Conservancy. Indeed, they even can transfer their accumulations of dead labor, and, by extension, dead nature, to The Nature Conservancy to tend the grave sites of that which they murdered cheeseburger by cheeseburger, BTU by BTU, freon molecule by freon molecule in their lethal mode of suburban living. Even more ironically, the hit men of these myriad murder-for-hire deals—or major corporations—also are solicited by the Conservancy to pony up land, capital, or donations to sustain this noble enterprise. Economy and environment are, of course, not incompatible, because this is the circuit of maggot and corpse, buzzard and body, grub and grave so common in today's postmodern ecology. Capital and Nature, the dead and the living, are incompatible, but capital has won: Nature is dead. All that is left is the zombie world of economies and environments, or the cash credits inside corporate ledgers for capital circulation and the ecological debits outside of corporate accounting charged off as externalities. Some still think capitalism has not yet defeated Nature, but they are deluded. Everything is environment now; nothing is Nature, except perhaps the last reaches of inner space and outer space where aquanauts and astronauts, riding high-tech robotic probes, have not yet come in peace, killing everything before them to then rest in peace.

This subtext in The Nature Conservancy's methods of operation is almost overpowering at times. It is "dead serious" about "its nonprofit business," namely, buying "the resting ground" of Nature among thousands of plots, dotting the nation from coast to coast on acreage "unpaved and as yet unplowed."[40] There TNC will guard its cemeteries for Nature, asking that no one bother the clients entombed inside. "For example," the Conservancy proclaims, "there's a sign in one preserve that says 'Rattlesnakes, Scorpions, Black Bear, Poison Oak/ARE PROTECTED/ DO NOT HARM OR DISTURB.'"[41] Like those historical societies that find, catalog, and guard old human graves in the memory of the nation's Puritan founding, Revolutionary War veterans, or Civil War fighters, The Nature Conservancy marks the death of Nature through the penance of an exacting inventory. It becomes a massive graves-registration effort, hoping "to create a huge, always up-to-date inventory of the rarest animals, plants, and natural places in each of the United States. Set up with state governments, these 'State Natural Heritage Programs' identify what's rare and what's threatened in each state: birds, butterflies, orchids, marshes, river systems, swamps, forests . . . and crane's nests."[42] Like the dead human past, dead Nature can be memorialized by saving what is rare and threatened, such as old tombstones or grave sites, through shrewd acts of scientific surveillance, state government, and land purchasing.

TNC prides itself on the national scope of its memorial sites. With fifty-nine chapters in fifty states, all one has to do is let his or her fingers do the walking through the phone book. Modeling itself on the rituals followed during those infrequent visits to old family graves, The Nature Conservancy always can be counted on to stand by ready and waiting to guide "you and yours to a nearby preserve where you're most welcome to walk along one of its paths, sit on one of the log benches, look about, and say to the youngster we hope will be with you, 'This will be here, as is, for *your* grandchildren.' Nice feeling."[43] Mother Nature is dead, but, thanks to TNC, her body will remain ever accessible to our great-grandchildren, who will be able to visit these perpetually maintained cemeteries to see the birds, orchids, swamps, forests, rivers, and marshes that are her many beautiful headstones. It will be a nice feeling knowing that our heirs might see such tokens of Nature behind a fence or wilderness from a log bench, in these terranean tombs, like the mortal remains of Lenin or Mao under glass, decades from now.

These aesthetic appeals, however, to preserve lands and scenery in keeping with the Conservancy's initial organizational agendas, just mystify the organization's more recent objectives of preserving biodiversity. Scenery provides legitimation, land creates a containment area, and rare ecosystems constitute storage sites for precious biogenetic information. Thus, these memorial parks for "nature conservancy" more importantly are becoming a network of cryogennic depots. Inside their boundaries, natural wetware accepts deposits as genome banks, accumulating bioplasmic memory on the hoof, at the roots, under the bark, and in the soil of Nature Conservancy protection actions. Nature is dead, but its environmental remains are put into a cryogenic statis until some future day when science and technology can bring the full productive potential out of them that escapes human development now. At that point, they too will be released from their frozen state to become the trade lands of tomorrow, as some snail, lichen, or bug is discovered to hold a cure for cancer or the common cold.

Under the guidance of Bob Jenkins's biodiversity plan, Nature has been transmogrified from the matter and space hoarded by the Ecologist's Union into informational codes and biospheric addresses archived by The Nature Conservancy. Plants and animals become more than endangered flowers or threatened fish; they become unknown and unexploited economic resources essential to human survival. "Of all the plants and animals we know on this earth," as one Conservancy supporter testifies, "only one in a hundred has been tested for possible benefit. And the species we have not even identified yet far outnumber those that we have. We destroy them before we discover them and determine how they might be useful."[44] Conservancy preserves, then, are biodiversity collection centers, allowing a free-enterprise-minded foundation to suspend their native flora and fauna in an ecologically correct deep freeze until scientists can assay the possible worth of the ninety-nine untested species out of each hundred banked in these preserves.[45]

Meanwhile, grizzly bears, bald eagles, and spotted owls provide high visibility entertainment value in its preserves for ecotourists, Conservancy members, and outdoor recreationists all seeking to enjoy such Edenic spaces. In "preserving Eden," the Conservancy more importantly is guarding the bioplasmic source codes that enable the wetware of life to recapitulate its existence in the timeless routines of birth, life, reproduction, and death.[46] Such riches can only be exploited slowly, but they cannot be developed at all unless today's unchecked consumption of everything everywhere is contained by Nature Conservancy protection actions bringing the world economy to an absolute zero point of inactivity in these Edenic expanses of the global environment.

4

Worldwatching at the Limits of Growth

Many ecological organizations are criticized by the mass media for being too "radical" in organizing environmentalist actions. By reminding us that pollution respects no borders, that the wind carries Chernobyl everywhere, that rain forests cut down in Brazil today may mean no cancer cures in Nebraska tomorrow, or that thinking globally requires acting locally, such mediagenic groups as Greenpeace, Earth First! and the Sierra Club have redefined the rhetorics of political radicalism, sites of global involvement, and visions of environmental activism in the imaginations of millions. These high-profile green groups may, however, only be bit players in media spectacles, which project televisual images that occlude far more "radical" styles of transnational environmentalism—like The Nature Conservancy at a local level or the Worldwatch Institute on the global level—busily at work elsewhere.

This chapter reconsiders the work of the Worldwatch Institute to question the idea of "sustainable development," which has become one of the world's most unquestioned environmental philosophies. The literature articulating this philosophy is vast and multidimensional, encompassing voices from the world's developed and underdeveloped economies. This brief consideration, however, will examine only one original source of sustainable development thinking, namely, the Worldwatch Institute as it has been led by Lester Brown. It stresses how the major continuing tension in sustainability debates between the operational objectives of "preserving Nature" and the practical ends of "maintaining the economy" has been reconciled by the Worldwatch Institute. In adopting a resource managerialist approach toward the environment, Brown and his fellow "worldwatchers" present visions of sustainability in nature preservation terms, while remaining firmly committed to maintaining economic growth (albeit with green concerns being given a starring role in all new economic initiatives).

The Worldwatch Institute: A First Glance

The credibility of the Worldwatch Institute, unlike Greenpeace, Earth First!, or even The Nature Conservancy, is very solidly grounded: big-time publishing ties, serious policymaker endorsements, and well-sited offices. A major commercial publisher, W. W. Norton, brings out its *State of the World* yearbook as well as the new environmental alert series called *Vital Signs*. The wide range of Worldwatch Institute publications, ranging from *World Watch* magazine (started in 1988), the *State of the World* annuals (first issued in 1984), and the Worldwatch Paper series (begun in 1975), has won widespread acceptance of Worldwatch Institute views in policy-making circles, the mass media, and scientific networks. This legitimacy is reflected in the organization's financial backing. One can survey any issue of the *State of the World* (1984–97) yearbooks for their acknowledgment of "the generous assistance" provided by a number of blue-ribbon corporate, international, and private foundations. As the 1991 *State of the World* annual affirms, "Worldwatch owes its existence to the largesse of a number of organizations that have supported our work over the years. Core funding for the *State of the World* series comes from the Rockefeller Brothers Trust and the Winthrop Rockefeller Trust."[1] Funding for the Worldwatch Institute's other research activities, in turn, comes from such granting agencies as the Geraldine R. Dodge, George Gund, William and Flora Hewlett, W. Alton Jones, William D. and Catherine T. MacArthur, Andrew W. Mellon, Curtis and Edith Munson, Edward John Noble, Public Welfare, Surdna, and Rockefeller foundations along with monies from the United Nations Population Fund.[2]

From the perspective of social impact and institutional accessibility, the Worldwatch Institute in many ways is one of the most visible environmental advocacy groups in the United States and the world today. The play given in the mass media to each new edition of the *State of the World* annual more or less affirms the message conveyed by each new volume's back cover copy, namely, that "in the absence of a comprehensive annual assessment by the United Nations or any national government, this book is now accorded semi-official status by national governments, UN agencies, and the international development community." Along with many other similar think tanks, political action committees, professional lobbyists, and interest group offices, the Worldwatch Institute's offices are headquartered on Massachusetts Avenue in Washington, D.C. Except perhaps by figures in the right-wing "wise use" conservation movement, all of these financial, organizational, and professional connections ordinarily would not be read as indications of "radicalism." Indeed, the voices raised by the Worldwatchers are extremely credible to many authorities in government, industry, and academe. As the Worldwatch Institute's own public relations relates to its publics, "when policymakers, journalists, heads of state, and concerned citizens need the best environmental information available, they turn to the Worldwatch Institute."[3] More than three hundred thousand copies of the *State of the World* annuals are sold worldwide, and it appears in more than twenty languages, including, as the Worldwatchers judge, "all of the major ones: Spanish, French, Chinese, Arabic, German, Italian, Japanese, Russian, and English," which is, as the Worldwatch Institute observes, more than even the global editions of the *Reader's Digest.*

These links to the relations of power are suggestive indicators well worth a closer look. In fact, the Worldwatch Institute's policy wonks are among the environmental movement's most radical (in the sense of posing fundamental changes), most transnational (in the sense of being avowedly global in outlook), and most environmentalist (in the sense of situating Nature in a seamless envelope of technical disciplinary surveillance) groups. The most interesting aspect of the Worldwatch Institute's self-image is its repeated affirmation of having "the definitive word on the condition of our planet—and on progress toward achieving a sustainable society."[4] This belief that it can, first, actually define these conditions, that it, second, already has done so authoritatively, and that heeding its definitive words are, third, the true measure of making progress toward "a sustainable society" clearly invites further examination. When we hear the Worldwatchers asserting that they, and virtually they alone, now offer "a vision of the global economy" that demands "a shift to a sustainable, and ultimately, more satisfying society," one must cautiously reconsider why they believe that their definitive design for transformation "confronts head-on the hard political choices necessary to make such a vision a reality."[5] Consequently, this chapter carefully rereads the work of the Worldwatch Institute—focusing on one recent publication, *Saving the Planet: How to Shape an Environmentally Sustainable Society* (1991) by Lester Brown, Christopher Flavin, and Sandra Postel— to develop a critique of its theory and practice.

Instituting the Worldwatch

Seeing the path of untrammeled industrial development as the cause of environmental crises, Brown, Flavin, and Postel attribute the prevailing faith in growth to "a narrow economic view of the world."[6] Any sense of constraint on further growth is cast by economics "in terms of inadequate demand growth rather than limits imposed by the earth's resources."[7] Ecologists, however, study the allegedly complex changing relationships of organisms with their environments, and, for them, "growth is confined by the parameters of the biosphere."[8] Ironically, for Brown, Flavin, and Postel, economists regard ecologists' concerns as "a minor subdiscipline of economics—to be 'internalized' in economic models and dealt with at the margins of economic planning," while, "to an ecologist, the economy is a narrow subset of the global ecosystem."[9] The discourse of dangers propagated by the Worldwatch Institute pushes for fusing ecology with economics to infuse environmental studies with instrumental rationality and defuse economics with ecological systems reasoning. Economic growth no longer can be divorced from "the natural systems and resources from which they ultimately derive," and any economic process that "undermines the global ecosystem cannot continue indefinitely."[10]

The Project of "Resource Managerialism"

The work of the Worldwatch Institute rearticulates the instrumental rationality of resource managerialism on a global scale in a transnationalized register. Resource managerialism is one very particular articulation of ecology. This is "ecology" as it has been constructed by modern nation-states, corporate capital, and scientific professional organizations. Although voices in favor of conservation can be found in Europe early in the nineteenth century, the real establishment of this particular approach to Nature as actual policy comes into being, first, with the closing of the open frontier in the American West during the 1880s and 1890s in the United States and, second, with the advent of the Second Industrial Revolution from the 1880s through the 1920s.[11] Whether one looks at John Muir's preservationist programs or Gifford Pinchot's conservationist codes, an awareness of modern industry's power to deplete natural resources, and hence the need for new protective arrangements for conserving resources or slowing their rate of exploitation, is well established by the early 1900s. President Theodore Roosevelt made these policies a cornerstone of his presidency. In 1907, for example, he organized the Governor's Conference to address this concern at the federal and state levels, inviting the participants to recognize that the natural endowments upon which "the welfare of this nation rests are becoming depleted, and in not a few cases, are already exhausted."[12]

Over the past nine decades, the fundamental premises of resource managerialism have changed significantly. On one level, they have become more formalized in bureaucratic applications and legal interpretations. Keying off of the managerial logic of the Second Industrial Revolution, which empowered technical experts (or engineers and scientists) on the shop floor, and professional managers (or corporate executives and financial officers) in the main office, resource managerialism has imposed corporate administrative frameworks on Nature in order to supply the world economy or provision national society with more natural resources through centralized state conservation programs.

To even construct the managerial problem in this fashion, Nature is reduced to a system of systems that can be dismantled, redesigned, and assembled anew to produce its many "resources" efficiently and in adequate amounts when and where needed in the modern marketplace. On a second level, during the 1970s and 1980s, resource managerialism transcended simple strategies of merely conserving available quantities of nonrenewable resources by moving toward more expansive programs of protecting various types of environmental quality and providing for new systems of renewable resource generation. Still, these shifts are not a major departure from the original premises of conservation. They only broaden the conceptual definitions of resources either being created from or conserved within Nature, while expanding the prerogatives of managerial authority to renew as well as conserve resources. By envisioning it as an elaborate system of systems, Nature can be continually tinkered with in this fashion to find new fields within its systematicities to rationalize, control, and exploit for the benefit of human beings in wealthy, powerful nation-states. Beautiful vistas, clean air, and fresh water are redefined as "resources" that should not be overconsumed or underproduced, and the managerial impulse easily can rise to this challenge by creating recreational settings, scenery, and ecosystem services as entitlements to be administered by the state for multiple use in the economy, society, and culture.

Although resource managerialism can be criticized on many levels, it has provisionally guaranteed some measure of limited protection to wilderness areas, animal species, and watercourses in the United States.[13] And, whatever its flaws, the attempt to extend the scope of its oversight to other regions of the world probably could have a similar impact. Resource managerialism directly confronts the existing cultural, economic, and social regime of transnational corporate capitalism with the fact that millions of Americans, as well as billions of other human beings, must be provisioned from the living things populating Earth's biosphere (the situation of all these other living things, of course, is usually ignored or reduced to an aesthetic question). And, if they are left unregulated, as history has shown, the existing corporate circuits of commodity production will degrade the biosphere to the point that all living things will not be able to renew themselves. Other ecological activists can fault resource managerialism, but few, if any, of them face these present-day realities as forthrightly in actual practice, largely because the prevailing regimes of state and corporate power, now assuming the forms of the "wise use" movement, often regard even this limited challenge as far too radical. Still, this record of "success" is not a license to ignore the flawed workings of resource managerialism. In fact, this forthright engagement with resource realities raises very serious questions, as the global tactics of such agencies as the Worldwatch Institute reveal.

Defining the Managerial Project

For the Worldwatchers, Nature is reduced to a cybernetic system of systems that reappears among the world's nation-states in those "four biological systems—forests, grasslands, fisheries, and croplands—which supply all of our food and much of the raw materials for industry, with the notable exceptions of fossil fuels and minerals."[14] In turn, the naturally self-regulating performance of these systems can be monitored in an analytical spreadsheet written in bioeconomic terms, and then judged in equations balancing constantly increasing human population, continually running base ecosystem outputs, and constrained potential for increasing ecosystem output given limits on throughput and input. When looking at these four systems, one must recognize:

> Each of these systems is fueled by photosynthesis, the process by which plants use solar energy to combine water and carbon dioxide to form carbohydrates. Indeed, this process for converting solar energy into biochemical energy supports all life on earth, including the 5.4 billion members of our species. Unless we manage these basic biological systems more intelligently than we now are, the earth will never meet the basic needs of 8 billion people.
>
> Photosynthesis is the common currency of biological systems, the yardstick by which their output can be aggregated and changes in their productivity measured. Although the estimated 41 percent of photosynthetic activity that takes place in the oceans supplies us with seafood, it is the 59 percent occurring on land that supports the world economy. And it is the loss of terrestrial photosynthesis as a result of environmental degradation that is undermining many national economies.[15]

Photosynthetic energy generation and accumulation is to be the accounting standard of environmentalizing discipline. It imposes upper limits on economic expansion; the earth is only so large. The 41 percent that is aquatic and marine as well as the 59 percent that is terrestrial actually are decreasing in magnitude and efficiency because of "environmental degradation." Partly localized within many national territories as bordered destruction and partly globalized all over the biosphere as transboundary pollution, the system of systems now needs global management, or a powerful, all-knowing worldwatch, to mind its environmental resources.

This imperative arises out of some dangerous convergent trends, as such Worldwatch bioeconomic accounting suggests:

40 percent of the earth's annual net primary production on land now goes directly to meet human needs or is indirectly used or destroyed by human activity—leaving 60 percent for the millions of other land-based species with which humans share the planet. While it took all of human history to reach this point, the share could double to 80 percent by 2030 if current rates of population growth continue; rising per capita consumption could shorten the doubling time considerably. Along the way, with people usurping an ever larger share of the earth's life-sustaining energy, natural systems will unravel faster.[16]

To avoid this collapse of energy and matter throughputs, human beings must stop rapid population growth, halt resource-wasting modes of production, and limit levels of material consumption. All of this requires a measure of surveillance and a degree of steering beyond the modern nation-state, but perhaps not beyond some postmodern worldwatch engaged in the task of equilibrating the "net primary production" of solar energy fixed by photosynthesis in the four systems. Worldwatching presumes to know how all of this actually works, and, in knowing it, to have mastered all of its economic/ecological implications. At the same time, these campaigns to redenominate knowledge in ecological categories are matched with an assault on the modern nation-state as a center of power.

These respecifications of some capillary conduits of control in Worldwatch Institute designs are captured in "a thumbnail sketch of a society that lives within its means" both ecologically and economically.[17] Although it is only a rough sketch, the Worldwatch Institute declares "how an environmentally sustainable society would look in the future," and then ticks off its managerial commands, directing us to key attributes:

- The world will be powered *not* by oil, coal, or natural gas, but by the *sun*
- Homes will be weather-tight and highly insulated, reducing the need for heating and cooling
- Kitchen appliances will be more than three times as efficient as those used today
- New light bulbs will reduce electric bills and last seven times as long as the light bulbs currently in use

- Aerodynamic four-passenger cars could get between seventy and ninety miles per gallon
- Communities will be more compact so shorter distances can easily be covered on foot or by bike
- Telecommunications could substitute for travel and shopping, cutting air pollution and fuel consumption
- Employment in a fossil-fuel economy will be replaced by new opportunities in home insulation, carpentry, wind prospecting, and solar architecture
- Waste reduction and recycling industries will replace the garbage collection and disposal companies of today
- Nutrients will be recycled and composting will become more commonplace
- The throwaway mentality will be replaced by a reuse and recycle ethic[18]

The tone and scope of these directives capture the spirit of the Worldwatch's managerial designs. Much of it is directed at finding alternative energy systems and technologies, or technological fixes, to solve problems caused by unsupervised cultural/economic/political systems.[19]

World power grids will shift to reliance on solar energy, not fossil fuels. Homes will be redesigned to save energy, and not signal style, status, or success first. Employment will change to pursue environmentalizing ends of sustainability, not economizing goals of growth. Waste reduction will displace waste production. Profligate consumerism will be replaced by preservationist austerities. To guard global carrying capacity, each subject must assume the less capacious carriage of disciplinary frugality. The world will come under watch, and the global watch will police its charges to dispose of their things and arrange their ends— in reengineered spaces using new energies at new jobs and leisures— around these managerial agendas. Still, at this level of analysis, the Worldwatch Institute does not do much more than stress the merits of a "green consumerism" that would have everyone live more ecologically at home, even though it would leave many regions of the corporate sphere free to pursue profits.[20]

The Worldwatch Institute admits that these managerial expectations will restructure the practices and qualities of subjectivity on a global scale, because "winning the battle to save the planet" depends not on changing "the values and behavior" of others but rather "depends on changing our own."[21] Encircled by the dictates of sustainability, the Worldwatchers expect humans to be swept up in a global mobilization to police the planet: "people everywhere will be involved directly: consumers trying to recycle their garbage, couples trying to decide whether to have a second child, and peasants trying to conserve their topsoil."[22] Most significantly, the Worldwatchers look beyond the modern nation-state, seeing it as an institution whose "importance may already have peaked." At the same time, they assert that dependence on global agencies such as the United Nations will grow "as environmental monitoring and enforcement powers are vested in international agencies."[23] Armed with environmental surveillance of population, land use, agriculture, energy use, and industry of this "worldwatch," the UN must become "worldaction," assuming an interventionist role, doing "for all people what national governments no longer can."[24]

These efforts to redraw state sovereignty in Worldwatch global accounting tallies obscurely express the resource managerialism project within a rhetoric of scientific technicalities. How to govern Nature, especially after its envelopment in resource managerial design gels in worldwatching disciplinary discourse, is not a purely administrative question turning on scientific "know-how." Rather, it is essentially and inescapably political. The discourses of worldwatching that rhetorically construct Nature, as suitable for governing contemporary society through particular knowledges, also must assign powers to new global governors and governments, who are granted writs of authority and made centers of organization in the Worldwatchers' definitions of managerial "who-can" and political "how-to."

Worldwatching at the Limits to Growth

As the Worldwatchers elaborate them, the discourses of sustainability center on advancing "sustainable development" rather than simply working to preserve Nature.[25] The World Commission on Environment and Development asserts: "in essence, sustainable development is a process of change in which the exploitation of resources, the direction of investments, the orientation of technological development, and institutional change are all in harmony and enhance both current and future potential to meet human needs and aspirations."[26] Here, as chapter 5 will suggest, the authors of *Our Common Future* buy into the worldwatching project as they engage in their elaborate exercises of green futurology; that is, sustainable development is a schema for organizing and administering "Our Future Commons" through a harmonious processing of peculiarly exploited resources, directed investment, oriented technologies, and changed institutions that enhance the prospects of needy humans aspiring to develop their potential. Underneath the enchanting green patina, sustainable development is about sustaining development as economically rationalized environment rather than the development of a sustaining ecology.

Localism and Globalism

The visibility and impact of such global commissions on the environment suggest to the Worldwatch Institute that the political landscape is on the brink of a major upheaval. Arguing that everyone recognizes "the inability of governments to protect their citizens from global environment threats," and that the importance of national governments "may have already peaked," the Worldwatch Institute sees power flowing up to global levels in "international agencies" and down to local levels in "citizens' environmental agencies"[27] since these political formations have been more receptive to the managerial exhortations of worldwatching discourses. It is not surprising that Brown and his collaborators are steering their relation of ecological truth to such political agencies as another effort to couple their systems of governing with the Worldwatch system of knowing. For the Worldwatchers, these policy maneuvers are already, in fact, happening, and, they are, in principle, necessary for real change. Armed with Worldwatch guidebooks, ordinary people are bringing "the struggle for a sustainable world" from "villages to the board rooms, from local town councils to the General Assembly in New York," because, as Brown, Flavin, and Postel assert, "it is only by bridging the vast chasm between the grassroots and international diplomacy that the pace of change can be sufficiently accelerated."[28]

At the same time, the sustainable society model of ecological critique arguably represents a considerable sophistication of the more crude neo-Malthusianism articulated in *The Limits to Growth*.[29] In its macroeconomic models of the constraints limiting future economic growth, the Club of Rome study misaggregated a great deal of questionable data, using several dubious assumptions in some very simple constructs that extrapolated the growth patterns of world economy prior to the 1973 OPEC oil shock into the indefinite future. Not anticipating any social learning or systemic crises that might change these behaviors, *The Limits to Growth* experts concluded: "If the present growth trends in world population, industrialization, pollution, food production, and resource depletion continue unchanged, the limits to growth on this planet will be reached sometime within the next one hundred years. The most probable result will be a rather sudden and uncontrollable decline in both population and industrial capacity."[30] The Worldwatch Institute adopts the global perspective used by "the limits to growth" school, but it assumes that there is a much more complex global system with many contradictory trends, working simultaneously in favor of conservation and waste, ecological care and antienvironmental neglect, social change and institutional inertia.[31] Instead of seeing a single inexorable push by the whole world up against uniform limits to growth in a single catastrophic collapse, worldwatching shows how ecological damage occurs daily on a piecemeal basis at varying rates of gradual degradation.

The Worldwatch critique is fairly straightforward. To satisfy the expanding needs of material consumption in both the developed and the developing countries, the world's ecosystems are being severely disrupted. In particular, three trends are moving dangerously out of the balance. First, many renewable resources are being depleted at a greater rate than they are being replenished. Second, many nonrenewable resources already have been consumed, while the remaining stocks are dispersed too widely to be collected and used efficiently. Third, pollutants are being discharged into some regions of the earth's soils, atmosphere, and oceans at levels beyond which they safely can be absorbed. Given these disruptions, entire regional ecosystems are being killed slowly through acidification, contamination, desertification, or soil salination. The organic basis of human life is being turned inadvertently into inorganic wastelands by human action. Because the really serious damage manifests itself slowly in many far-flung regions on a sector-by-sector basis, it often can be ignored or denied.

The global inventory conducted in each *State of the World* annual confirms these observations by refusing to ignore or deny the implications of these trends. With very little variation over the years, every new volume of this yearbook brings several chapter by chapter critiques of how energy is wasted, materials are misused, cropland is eroding, cities are inefficient, nuclear power is dangerous, or consumers are greedy. Other chapters point out how a more sustainable use of resources in a renewed culture, reshaped economy, and reformed society might work. The emphasis is not on finding the physical limits to growth to predict a general social collapse, but rather on identifying the ecological basis of sustainability to prevent complete environmental and social crisis.

Sustainable Societies

The notion of "sustainable development," as Redclift observes, "was used in the Cocoyoc declaration on the environment and development in the early 1970s. Since then it has become the trademark of international organizations dedicated to achieving environmentally benign or beneficial development."[32] Brown and the Worldwatch Institute have made it a cornerstone of their environmental vision. The "sustainable society," in Brown's analysis, however, is defined mainly as the opposite of the present "unsustainable society," namely, "population size will more or less be stationary, energy will be used far more efficiently, and the economy will be fueled largely with renewable sources of energy."[33] The present forms of everyday life must be transformed to match these new modes of production. The model is vague, but it leads to some very significant changes: "the transition to a sustainable society promises to reshape diets, the distribution of population, and modes of transportation. It seems likely to alter rural-urban relationships within countries and the competitive position of national economies in the world market. Then too, a sustainable society will require labor force skills markedly different from those of the current oil-based economy."[34]

Brown does not present a workable program for making this transition to a sustainable society. The changes he identifies are instead policy imperatives derived from statistical analyses of long-term tendencies in energy use and resource consumption for "the world" from the 1980s through the early twenty-first century. Because oil, coal, gas, and nuclear power will run out and/or pose tremendous ecological side effects, a renewable energy base will arise out of need. With this base of renewable energy rather than fossil or nuclear fuels, the entire remaining superstructure of society allegedly will change, leading to a sustainable transportation system, a resurgent agriculture, new industries and new jobs, new cities, simpler lifestyles among affluent peoples, greater local self-reliance, and more autonomy in the Third World.[35]

The transition is urgent because of "the tendency of the negative trends to reinforce themselves."[36] And, since the basic change in values that would make such a transition most feasible is occurring very slowly, Brown advocates embracing market forces plus using financial carrots and sticks to move contemporary capitalism toward the practices of sustainable society. Through subsidies, high user fees, taxation schemes, and government regulation, Brown sees the profit motive, or individual households, firms, and nations recognizing the savings possible in renewable energy, forcing modern societies away from "growth" and over to "sustainability." Still, he ultimately concedes that "the greater the need for economic and social change, the greater the need for leadership to guide the process."[37] Seeing a dearth of good national political leadership everywhere, Brown instead claims that most of the available talents "actively promoting the transition to a sustainable society are local."[38] In thousands of local communities, he claims that ordinary people are already taking the essential steps, launching innovative recycling, energy conservation, population planning, and land conservation programs. Again, Brown opts for people power over state power: "everyone will have the opportunity not only to participate in the transition but to help lead it."[39]

Even so, the Worldwatch approach to corporate capital reflects an ambivalence that undercuts its critique. Brown notes that "modern corporations exercise enormous influence over our daily lives," simply by being concentrations of economic power and technical resources that rival those of many smaller nation-states.[40] As such, "corporations are in a position to either facilitate or slow the transition to a sustainable society."[41] He recounts how Standard Oil, General Motors, and Firestone colluded to dismantle urban rail-based mass transit systems in American cities after 1945. Likewise, he observes how Detroit refused to produce small cars in the 1970s oil crisis because they meant small profits, and how corporate utilities continued to construct huge, unnecessary new plants instead of funding conservation and efficiency. Nevertheless, he finds redeeming social value in corporations because they can bring "their impressive R & D capacity to bear on the transition";[42] and, "to the extent that new techniques and technologies can play a role in the transition—be they tree-breeding techniques or the development or more energy-efficient industrial processes—companies can begin to assume social responsibilities commensurate with their size and wealth."[43] Thus, corporations "can help" to develop renewable energy, sustainable societies and simpler lifestyles.

Beyond touting vague potentials for profit, however, Brown is very unclear about why corporations might cooperate: "those companies that understand the economic transition and plan accordingly will benefit. Those that fail to anticipate the changes in prospect may find themselves foundering."[44] The companies that actually fail to plan or resist the transition by advancing the wrong kind of R & D are essentially ignored. Yet, this antiecological pattern has been the trend for nearly fifty years of the post-1945 ecological disaster. Companies can persist in producing gas-guzzlers and pollutants, recognizing that buyers will gladly pay the punitive taxes to still enjoy their "consumer goods." Brown supports a transition strategy that creates a market in pollutants that would sustain the current industrial regime and its ecological irresponsibility by commodifying previously ignored externalities. Facing the established and entrenched powers of corporations, then, is really not what Worldwatchers will do. Instead, the real challenge of fighting for a "sustainable society" in the Worldwatchers' programs turns more toward endorsing "voluntary simplicity" and "conspicuous frugality" for consumers, while leaving the corporate world to respond to these new "market forces."[45]

Although Brown and his associates never admit that the world has slipped into the era of postmodernity, their practices do lend support to Jameson's claims that postmodern times begin "after the modernization process is complete and nature is gone for good."[46] For the most part, the Worldwatchers assume that the existing technology base of ad-

vanced industrial society is what the world must work with at this juncture. The modernization process has produced this array of technologies
with all of their negative and positive benefits, but, at the time, there
will be no major new breakthroughs or improvements. The advances of
modernization are over, but redistributing the impact of its benefits and
costs has not yet really begun on a rational basis. The Worldwatch project, then, is dedicated to surveying costs and benefits on a global scale
in order to make the administrative interventions needed to provide the
maximum sustainable benefits to the greatest number and to limit ecological costs to the lowest level for all. To do this, worldwatching essentially acts as if Nature is gone for good, because the global environment
that it now watches is not composed of the wild, unknown powers of
Nature. It is instead an ensemble of ecological systems, requiring human
managerial oversight, administrative intervention, and organizational
containment.

Encircled by such circuits of ecological alarm, the Worldwatch Institute's many sustainability discourses tell us that today's allegedly
unsustainable growth rates in unstable environments can be corrected
simply by allowing all industrial metabolisms to be disassembled, re-
combined, and subjected to the disciplinary designs of its expert management. Enveloped in such interpretative frames, any environment can
be redirected to fulfill the ends of other economic scripts, managerial directives, and administrative writs denominated in sustainability values.
Sustainability, as a goal of governmentality, also engenders its own forms
of "environmentality," which would embed alternative instrumental
rationalities—beyond or beneath those of pure market calculation—in
the policing of ecological spaces.[47]

This meaning for sustainable development in Worldwatch discourse reframes it in the practices of technoscientific power/knowledge.
One can argue that the modern regime of biopower formation described
by Foucault in early modern states was not especially attentive to the
role of Nature in the equations of biopolitics.[48] The controlled tactics of
inserting human bodies into the machineries of industrial and agricultural production as part and parcel of strategically adjusting the growth
of human populations to the development of industrial capitalism, however, did generate systems of biopower. Under such regimes, power/

knowledge systems bring "life and its mechanisms into the realm of ex-plicit calculations," making the manifold disciplines of knowledge and discourses of power into a new sort of productive agency as part of the "transformation of human life."[49] Once this threshold was crossed, some observers began to recognize how the environmental interactions of human economics, politics, and technologies continually placed all human beings' existence as living beings into question.

Foucault might be read as dividing the environment into two sepa-rate but interpenetrating spheres of action: the biological and the histori-cal. For most of human history, the biological dimension, or forces of Nature acting through disease and famine, dominated human existence with the ever-present menace of death. Developments in agricultural technologies as well as hygiene and health techniques, however, gradu-ally provided some relief from starvation and plague by the end of the eighteenth century. As a result, the historical dimension began to grow in importance along with "the development of the different fields of knowledge concerned with life in general, the improvement of agri-cultural techniques, and the observations and measures relative to man's life and survival contributed to this relaxation: a relative control over life averted some of the imminent risks of death."[50] The work of the Worldwatch Institute acknowledges how "the historical" then begins to envelope, circumscribe, or surround "the biological," creating interlock-ing disciplinary expanses for "the environmental" to be watched, man-aged, controlled. And, these environmentalized settings quickly domi-nate all forms of concrete human reality: "in the space of movement thus conquered, and broadening and organizing that space, methods of power and knowledge assumed responsibility for the life processes and undertook to control and modify them."[51] Although Foucault does not explicitly define these spaces, methods, and knowledges as such as being "environmental," these governmentalizing maneuvers might be seen as the origin of many disciplinary projects, which all feed into environ-mentalization. As biological life is refracted through economic, political, and technological existence, "the facts of life" pass into fields of control for disciplines of ecoknowledge and spheres of intervention for their management as geopower at various institutional sites, such as the Worldwatch Institute.

In the disciplines of worldwatching, the raw powers of "Nature" are erased, leaving behind only the systematic patterns of "the global environment." They represent the world as a closed totality through "eco-knowledges" that will disclose its logics, interconnections, and operations as "geopower" seen by correctly informed analysts. This construction also reveals the Worldwatchers' actual relation to the world as such; they sit above, outside, beyond the sites of greatest crisis as analytical advocates, who now are powerless but also seek empowerment through their reformist advocacy of particular strategies of change. Worldwatching, then, is perhaps the necessary and expected outcome of postimperial transnationalism. It extends and elaborates a pluralist model of countervailing political organizations on a global level from the home bases of transnational business. As policy wonks and scientific experts, they have mobilized various scientific communities to participate in monitoring the environmental crisis and set policy agendas. Yet, the watchdog function of environmentalism is globalized without a serious forum or effective mechanism for exerting political pressure. Sometime in the future, if and when this Worldwatch lobbying campaign works, it hopes to effect real change through local governments and the agencies of the United Nations. In the meantime, it centers itself in Washington, settling for lobbying the American state and multilateral aid agencies as the next best pressure point.

All Along the Watchtower: The Final Look

The Worldwatch Institute casts the existing world economy and ecology as a single integrated system, seeking to appraise "the relationship between ourselves and the natural systems and resources on which we depend."[52] In considering the existing world, the Worldwatchers specifically focus on "the links between human activity and environmental degradation."[53] After identifying the links, worldwatching investigates how human energy use, population growth, pollution, and resource exploitation promote various kinds of environmental damage. As an inventory of crises, detailing the depth and breadth of environmental destruction around the globe, worldwatching performs an important information-gathering and news-disseminating function, but it really fails to confront the basic causes of these crises.

In the *State of the World* (1988), for example, the planet is presented as undergoing "a physical examination" to check "its vital signs."[54] As part of this physical, the Worldwatch Institute finds forests dying, deserts expanding, soils eroding, the atmosphere deteriorating, lakes acidifying, energy stocks failing, and human populations rapidly expanding, even as many plant and animal species are dying out. It correctly and clearly identifies many problems. Still, it analytically treats them as separate maladies, reflecting different threats against "the world's life-support systems."[55] The basic logic of commodification and exchange that causes ecological destruction in the core economies simply is neglected.

In the final analysis, then, worldwatching is little more than a global doctrine for applying a range management philosophy to a wide array of human communities within a diverse but interrelated set of degraded ecological ranges. Different managerial strategies are needed for various human groups in core, semiperipheral, and peripheral ranges. For the developed, their existing forms of material culture, industrial infrastructure, and economic exchange basically will be retained, but the code of sustainability requires that patterns of energy, resource, and land use be redesigned to allow society to follow essentially many of the same paths. Through intensified campaigns of energy conservation, resource recycling, and product reengineering, the "coreness" of highly developed regions might be sustained with the right leadership—highly motivated local and international agencies working closely with these Worldwatch "range managers"—through finding greater efficiencies in the existing capitalist economy by reducing both production and consumption.

Yet, the Worldwatch Institute's resource managerialist analysis also tells us that our biotic range simply does not have the depth of resource renewal or the breadth of carrying capacity to bring everyone on the planet up to core levels of development. Hence, sustainability theory directs worldwatching range managers, based in core regions of developed societies, to work toward holding constant (at an intermediate range of technical, economic, and social development) much of the semiperiphery and periphery. Because these economic regions have much larger populations whose potential rates of core-level material consumption, if ever attained, would overwhelm the world's ecological balance, they must balance the world's ecology by accepting those virtues that the core sees in sustainability. To the extent that such divisions still make any sense, the world that is being "watched" is largely that in the Second, Third, and Fourth Worlds. In turn, the world that is being "watched out for" is essentially the First World inasmuch as it represents the high technology and high material consumption core of the current global economy. Sustainability, then, will mean different things in these different worlds, because they all start at different levels of already attained material prosperity that either can be or will need to be sustained to maintain environmental balances.

The health of global populations as well as the survival of the planet itself necessitate that a green spreadsheet be draped over Nature, generating an elaborate ecomarket of global reach and scope. Hovering over the world in a scientifically centered surveillance of health, disease, poverty, wealth, employment, and joblessness, Brown, Flavin, and Postel declare that "the once separate issues of environment and development are now inextricably linked."[56] Indeed, they are, at least, in the discourses of Worldwatch Institute disciplines, which then survey this envelopment of Nature-in-crisis by auditing levels of topsoil depletion, air pollution, acid rain, global warming, ozone destruction, water pollution, forest reduction, and species extinction.

Worldwatching engages in a continuous global surveillance sweep, searching over patterns of energy use, artifact manufacture, food production, shelter construction, waste management, and urban design for technical, managerial, and economic inefficiencies. Once these searches are concluded, the results indicate, as the Worldwatch Institute reads them, there is a need for a permanent perestroika, or an ongoing, unending restructuring of everything artificial that extracts matter and energy from Nature in order to more rightly dispose of things and more conveniently arrange ends.[57] Here, once again, the peculiar environmental project of the Worldwatch Institute shows its hand. The goal of sustainability, on one level, has many laudable intentions driving its designs; yet, on another level, its discursive framing, its intellectual articulation, and its action planning already provide a power formation, a discursive center, and a rhetorical foundation to empower worldwatchers to stand watch over everything and everyone else in the name of their resource managerialism to attain bioeconomic efficiency.

5

Environmental Emulations: Terraforming Technologies and the Tourist Trade at Biosphere 2

On September 17, 1994, a second shakedown crew of seven Biospherians left their living quarters in Biosphere 2, ending abruptly their planned six-month session of experiments. Intended originally to secure some loose ends exposed by the first mission of Biosphere 2, which ran from September 1991 to September 1993 with a crew of eight in the glare of international media spotlights, this far more low-key test was called to a dead stop by Edward P. Bass, the Texas billionaire who has backed the project financially from the start. Wishing to cut all ties to the original management team and controversial design philosophy that launched Biosphere 2 ten years earlier, Bass also used this occasion to announce a new alliance between the Biosphere 2 environmental research facility and Columbia University's Lamont-Doherty Earth Observatory.[1] All of these moves followed from measures taken by Bass during April 1994, when he called in local marshals to oust the project's original managers in a dispute over, first, how the Biosphere 2 had been administered during its original shakedown mission and, second, what scientific orientation the facility should adopt in its work.[2]

Despite these developments, it is hard not to be impressed by the scope of the original Biosphere 2 experiment. More than ten years of planning, building, and testing have paid off in the development of an awesome architectural structure and an intriguing environmental exercise, which had been dedicated to "reproducing Planet Earth" in order to learn more about nature's ecologies. The undertaking never escaped controversy, and critics of the project—both from within the scientific community and among the general public—have been numerous and vociferous. Yet, they have tended to focus either on the project's founders for their self-important celebration of Biosphere 2's dubious New Age philosophical project or the questionable research agendas at the core of the experiment for lacking a certain scientific rigor, replicability, or robustness. Such criticism is well taken, but most considerations of Biosphere 2 are still far too accepting of its professed purpose, namely, allowing scientific technology full play at "reproducing Planet Earth."

This chapter problematizes Biosphere 2's original mission, questioning, first, how its reproduction of terrestrial ecology actually works and, second, the nature of the "Planet Earth" being reproduced. At the end of the day, Biosphere 2 appears in many ways to be an attempt to replicate technologically a naive anthropocentrism as the fundamental design rule for operating the earth's biosphere rather than a new collective defense technology for guarding Nature from further ecological degradation. Consequently, this reexamination explores some less openly discussed aspects of the biophysics, political economy, and history behind the original Biosphere 2 mission to cast new doubts over its various ecological commitments.

The Origins of Biosphere 2

Biosphere 2 was launched during 1983 by eight members of a group known as the Decisions Team. John Allen, Marie Allen, Margret Augustine, Edward P. Bass, William Dempster, Robert Hahn, Kathelin Hoffman, and Mark Nelson had been working together on various assignments since 1974, mostly through the auspices of the Institute for Ecotechnics.[3] Although it was often inchoate, the basic agenda of the Institute for Ecotechnics was to exploit the possible synergies of integrating ecology with technics to integrate more rationally the technosphere with the earth's biosphere. In many ways, it represented a fusion of New Age Gaia consciousness with hard-nosed environmental engineering. Bass, a wealthy Texan with his own venture capital firm, Decisions Investment, proposed a joint venture between his company and the Decisions Team to be called Space Biospheres Ventures (SBV). Bass assumed the post of chairman of the board, Augustine became president and CEO, Dempster served as the chief engineering designer, Nelson was assigned responsibility for space applications, and John Allen was appointed executive chairman. Augustine, who also is a professional architect and managing director of Synergetic Architecture and Biotechnic Design in London, would design Biosphere 2 in collaboration with her colleague Philip Hawes.[4]

After an intensive search for a site to build Biosphere 2, SBV settled on southern Arizona, and the Decisions Team found a promising location twenty-five miles north of the city of Tucson near the small town of Oracle. The Sunspace Ranch, as the Decisions Team tagged the facility, encompassed more than two thousand acres in the foothills of the Santa Catalina Mountains above Canyon del Oro, and featured excellent conference and lodging facilities since it had been serving as the site of the Motorola Corporation's Executive Training Center. SBV moved onto the grounds by July 1984, and began designing a small-scale test module to simulate the larger Biosphere 2 ecosystem. In 1985, Augustine specified that this facility should test all of Biosphere 2's major structural components and life science systems. By January 1987, the experiments on the test module had begun, and they continued through 1989 with John Allen and Abigail Alling, SBV's associate director for development and a prospective Biospherian for Biosphere 2, occupying it during these experiments as the human participants.

To design the full-scale Biosphere 2 system, SBV also retained the services of many different ecological, engineering, and environmental management experts. At the University of Arizona, Carl Hodges and his Environmental Research Lab joined the team to design the agricultural and engineering systems needed to maintain the new biosphere. Walter Adney from the Marine Systems Laboratory at the Smithsonian Institution designed the ocean and marsh biomes. Ghillean Prance, the director of Kew Gardens in the United Kingdom, helped to organize the rain forest biomes. Tony Burgess and Peter Warshall, both of the University of Arizona, agreed to help create the desert and savanna biomes of Biosphere 2. Clair Folsome of the University of Hawaii, who pioneered experiments with closed biospheres in the 1960s, assisted SBV with soil biology and chemistry.

Commercial interests, as the involvement of Space Biospheres Ventures illustrates, always have driven the Biosphere 2 experiment. And its profit potential has been closely tied to the space program. John Allen, one of Biosphere 2's founders, suggests in his *Biosphere 2: The Human Experiment*:

> In 1984 NASA plans called for Space Station *Freedom* to be in orbit in 1992. Space Biospheres Ventures drove to get Biosphere 2 built and into operation by that date, anticipating the possibility of putting the first small space life system into orbit by 1995. Venture capital was raised on the assumption that marketable technology would be developed, which would offer practical solutions to specific problems of pollution control and environmental management on Earth. The Biosphere 2 project would be not just a matter of science and technology, important as they were, but also one of appropriate finance, management, and product development.[5]

Biosphere 2's experimental emulations of Earth ecology, therefore, were aimed at generating marketable technologies to reconfigure many terrestrial human/environment material interfaces through the industries of pollution control and environmental management. At the same time, science and technology were to be mobilized by this enterprise in the 1980s to sell a small space life system to NASA for launching into earth orbit by 1995. These twin markets—sales to NASA and environmental management industries—were the key targets for Space Biospheres Ventures.

After being sealed in September 1991, the Biosphere 2 experiment unfolded over the course of the next twenty-four months as four men and four women lived at the top of its food chains and tended its technospheric infrastructures to test the viability of the Biosphere concept.[6] From the outset, it was plagued by mishaps—minor medical traumas, serious oxygen shortages, extreme temperature variations, carbon dioxide buildups, unplanned animal occupations, and insufficient food production—that created considerable friction among its managers during those first two years. The first Biospherians emerged in late 1993 declaring the experiment to be a success, but few outsiders were so positive. After promising to keep the original eight-member crew sealed up for two years without interruption, Biosphere 2 had to be opened almost immediately to respond to an emergency surgery situation and to replace some supplies. An unusually cool and cloudy winter limited the productivity of Biosphere 2's food systems, forcing the Biospherians to go hungry and lose weight. During the first experiment at producing food within its fully enclosed ecosystems, the buildup of excessive levels of carbon dioxide in the Biosphere's structure was an even larger problem. Some animal species completely died off (for example, all of the honey bees), and oxygen had to be pumped into the structure to mitigate the dangerously high levels of carbon dioxide.[7]

Because of disagreements over how these problems were dealt with internally as well as publicized externally, Bass forced out most of the original managers from the Decisions Team in April 1994 and installed Stephen K. Bannon, a Harvard Business School graduate who made his managerial mark in similar "knowledge-based companies," to serve as the chief executive officer of SBV. Bannon, in turn, tapped Bruno D. V. Marino from the Department of Earth and Planetary Sciences at Harvard University to take over as the science director of Biosphere 2 during August 1994. Marino and several other researchers also have begun reviewing the scientific potential of Biosphere 2's enclosed ecosystem, which may well be best exploited by taking human beings out of the experiment's simulated environment.[8] The sudden cessation of the second mission by a new Biospherian team in September 1994 simply underscored this complete change in operational focus.

The Biosphere 2 structure encloses 3.2 acres of land in the largest, most tightly sealed, fully closed environmental system on Earth. Modeled on Earth, or Biosphere 1, this artificial system brings together around 3,800 species of plants and animals as well as eight humans into the ecologies of seven basic biomes—human habitat, intensive agriculture areas, marsh, savanna, tropical rain forest, desert, and a twenty-five-foot-deep ocean and coral reef.[9] These biomes are arrayed inside a vast tetraheral space frame and glass superstructure. Its basic geodesic motif continues through five distinct spaces, dominated by two large, flat-topped, three-tiered pyramids—one to the north, one to the south. Between them, a lower but also pyramidally arrayed hall connects the two pyramids as well as a smaller hall topped by three abutting arched vaults blending into three squat domed turrets and a short geodesic dome-capped tower. Simultaneously suggesting visions of a NASA moon base, a counterculture commune, a Mesopotamian ziggurat, a climatronic greenhouse, a Mayan ruin, or a sci-fi hideout, Biosphere 2 is an architecturally striking edifice surrounded by its allied otherworld-looking support structures and the festival marketplace buildings of its entertainment centers: a hands-on science fair, a theater, gift shops, sidewalk cafés, a juice and coffee bar.

The two big pyramids of the Biosphere 2 complex contain five of the seven biomes it encloses. In the north pyramid, there is the rain forest, which flows into the ocean and savanna biomes, as well as the freshwater and saltwater marshes. These zones, in turn, blend into the thorn scrub and desert areas in the south pyramid. The arched vaults contain the intensive agriculture biome, and the human habitat occupies the smaller geodesic dome modules along the north face of the intensive agriculture biome. These two areas constitute a simulated microcity sitting atop the complex food chains stretched through the other biomes into the intensive agriculture plots.[10] The visible architectural reliefs of Biosphere 2 also hide most of the mechanical infrastructure of fans, blowers, pumps, piping, and motors needed to make this closed environmental system operate. Housed mostly in the basement, these components feed and sustain the air chambers, composters, water tanks, dryers, condensation chambers, and controllers used to keep air and water moving, plants and animals alive, temperature and humidity constant.[11]

Technosphere as Ecosphere

In Biosphere 1, the biosphere of first nature provides the necessities required for sustaining human activities in the technosphere of second nature that each established society fabricates as an integral part of its human community. Biosphere 2 attempted to reproduce these ecological relationships in the integration of its seven distinct biomes. The basic six biomes—or the ocean/marsh/desert/savanna/rain forest/intensive agriculture regions—are all linked together to energize the seventh, or the human habitat, poised at the top of their tightly coupled food chains. This manifest environmentalist emulation of Biosphere 1, however, muddles the mechanistic material realities resting beneath and beside Biosphere 2. Biosphere 2's simulation of Biosphere 1 actually must reverse the relations of energy, matter, and information exchange prevailing on the earth; here, the Biosphere directly depends on an elaborate technosphere to sustain itself by pulling resources into it from Biosphere 1.[12] The biophysics of the planet itself are partially simulated by 200 motors, 120 pumps, 60 fans, and more than 50 miles of pipe. Although the system is partly powered by sunlight, the complex also needs steady inputs of electricity as well as hot and cold water supplies coming from outside of its internal loops (these power inputs are provided by natural gas-powered generators).[13]

To stabilize its artificial metabolic systems, Biosphere 2 also needs, first, its atmosphere to be processed through two vast "lung" structures that regulate its internal air pressure and preserve the structural integrity of its space frame and glass housing and, second, the heat exchange provided by three cooling towers that vent excess heat. At some point in the future, the environmental services generated by these active systems might be provided by much more passive substitutes. However, that point is quite far off. Consequently, Biosphere 2 remains an essentially industrial apparatus, integrating machinery, computers, chemicals, plants, animals, and soils into a closely coupled cybernetic mechanism that produces an unstable but highly marketable product—a simulation of Earth's total environment.[14]

Simulating Nature, however, is an ambiguous project. It can cut several ways. One might attempt to imitate some particular Earth environment by importing every element of its naturally occurring constituents into a new site devoted to the imitation. Or one could select the key elements in a particular natural environment and import as many representatives of these naturally occurring biota as possible into the imitation. Or one could reduce the central functions in a particular natural environment by identifying only the absolutely essential operators in its ecosystemic transformations, manufacturing an analogue of its interactions, by finding that minimally functional subset of biotic actors needed to make this real ecological system work. Or, finally, one could fabricate an essentially new synthetic ecosystem by mixing and matching various components—soils, plants, and animals—from a wide range of naturally occurring ecosystems to model a simulation, or a copy, of Nature, for which there really is no original. Biosphere 2 plainly takes the latter course, in each of its individual biomes as well as in all of the integrative links knitting together their collective ensemble of operations.

The biomes appear to be "real," but in actuality they are only impressionistically modeled, for example, on very vaguely conceptualized geographic regions, such as the Brazilian rain forest, Chilean coastal desert, or African savanna, which fuse nation-state referents with ecological life zone labels to identify certain environments. The specific species of plants and animals deployed to assist the emulation of these real ecosystems are arrayed artificially in combinations that occur nowhere naturally in Nature, because the planners have pulled plant biomasses and animal actors into their model on the basis of their potential ecological performance as bionic agencies and not their actual natural occurrence as Earth life-forms. Actually, Biosphere 2 is a unique ecoengineering project that reduces natural life-forms to their biotic/biophysical operability in order to reintegrate them in new synthetic ecosystems that can, in turn, develop only in the artificial spaces of this biospheric laboratory. Here, "Nature" is not Nature, but rather something that has been digitally sampled, botanically colorized, zoologically compressed, and ecologically scanned into a biospheric simulation of itself that could not and would not exist without the engineering needed to stage this odd ecological experiment.

Biosphere 2's Earth is a robotic Earth, designed to imitate the original by going beyond it in a series of rationalizing re-formations of its environmental logics. The Earth environment in these frameworks of analysis simply becomes "the biological life support system" whose internal cybernetic mechanisms can be reduced through instrumental reasoning to bioregenerative technologies. Biosphere 2's emulation of Biosphere 1 provides a multipurpose test bed for redesigning the earth as a series of bionic environmental technologies for future Biospheres 3, 4, 5 . . . N, which will be sent to the moon, shot into deep space, built in the polar regions, or sunk under the oceans.

In Biosphere 1, one rarely thinks about the biophysical inputs he or she takes from the environment every day. When it is viewed as a biological life support system, however, it becomes clear that

> approximately 0.6 Kg food, 0.9 Kg oxygen, 1.8 Kg of drinking water, 2.3 Kg of sanitary water and 16 Kg of domestic water for a total of 22 Kg per day, or some 45–50 pounds are required to provide life support for each person per every day in an artificial life support system.

> Thus, in the course of a year, the average person consumes three times his body weight in food, four times his weight in oxygen, and eight times his weight in drinking water.[15]

With such calculations, a human being's biophysical carrying requirements and the earth's biophysical carrying capacities are reduced to differentiable but integral engineering functions. Caught in the grids of scientific surveillance, the ecological interface of human organisms and biological environments can be transformed into technological design criteria either "to sustain human life in space on a permanent and evolving basis" or to exploit "the commercial opportunities and historic importance for such spinoffs" in bioregenerative technologies.[16]

A child's terrarium also tries to imitate the ecology of Earth, but this undertaking at Biosphere 2 is more than imitation—it is refunctioning, rerationalizing, redesigning what is taken to be the original planet in a new artificial world with all of the Biosphere 1's existing anthropogenic reengineering already embedded in its new fully enclosed ecosphere. Biosphere 2 shows how extensively Biosphere 1 is not now, or at least is no longer, what might be identified as Biosphere 0—or Earth, as it issued from Nature, God, or the Cosmos prior to humanity's evolutionary emergence. A cosmogenic, theogenic, or geogenic Earth perhaps existed before human beings or, at most, before the Neolithic revolution. But, for nearly ten thousand years, the earth has been increasingly reworked as zones of anthropogenic change, so "Humanity" and "Nature" have been a coevolutionary couplet for many millennia in Biosphere 1.

Until 1500, these anthropogenic changes were mostly modest and localized. Since 1800, they have become much more considerable and globalized. And, after 1950, they are important forces in the earth's biosphere. Yet, their very pervasiveness now leads many groups, such as Space Biospheres Ventures, to search for even more efficient expressions of anthropogenic change, seeing that the ordinary outcome of human technological activity often is not optimal for either "the environment" or "the human organisms" caught in their dyadic interaction. Hence, Biosphere 2 experiments in imaginatively reengineering basic environmental design by linking ecological biotic operators into new thermodynamic formulas built into a closed environmental machine. In making these methodological moves, the earth rhetorically is reduced to a self-contained biospheric apparatus or one huge process technology used to distill a complex product: a habitable environment.

Biosphere 2, then, has been a confused tangle from its inception. Organized as a scientific simulation of the earth, it has been run as another roadside attraction for the greater Tucson region's tourist industry. Designed to be a credible scientific project, it has operated throughout its brief history as a media event and technoscience soap opera. Funded first as a private venture capital exercise, it survived by huckstering other products to the consuming public—science shows, motel facilities, restaurant meals, flashy T-shirts—in order to keep its doors open. The rhetoric of the visitors' brochures captures the dual personality of Biosphere 2. On one level, as Space Biospheres Ventures hopes, the many and varied "commercial applications of the Biosphere 2 project's technology and research are far-reaching."[17] But, on another level, as the Biosphere 2 ecological theme park pretends, "the knowledge we are gaining from studying the complex interactions of the Earth, atmosphere, and life is the real reward."[18] Nonetheless, the ultimate agendas of the original group from the Decisions Team and the Institute for Ecotechnics blended dubious scientific practices and New Age philosophies of Gaia consciousness in an infotainment package that seriously undermined the professional scientific credibility of Biosphere 2. As Cary Mitchell, a NASA-funded ecological researcher at Purdue University, observed about Biosphere 2's original project, "it's fascinating, but there has been no real hypothesis testing or peer-reviewed research"[19] underpinning most of the facility's experiments.

The Commodification of Ecology

The ecology that has been packaged and sold at Biosphere 2 is a peculiar sort of corporate ecology—one reverse-engineered by Space Biosphere Ventures to optimize its thermodynamic efficiencies for maximal output with minimal input in extraterrestrial settings. The biomes of this biosphere, unlike those of the earth, are rigidly homogenized, stabilized, and centralized. Everything is monitored by a network of computers fed with endless streams of information from remote sensors. There are no discontinuities or disruptions in the pace of this planetary emulation;

everything is under control on its specially preselected tracks of an engineered environmentality. The organizational aspiration of this enterprise is to commodify environments as they work on a global scale by reducing them to fungible micrological applications. To construct a Biosphere 2 is to engage already in biosphering—a service that might be bought and sold like any other. Biosphere 2 does the Worldwatch Institute one better by fabricating its own little world under glass as if it were a watch to be worked, wound, and worn as a regulatory mechanism to control human affairs in Biosphere 1 in accord with its new meters of environmental stability and security.

This sort of corporate ecology also appears to be a product intent on selling people some of what they once had at no cost after other corporate ecologies degrade the environment so completely that Biosphere 1 no longer can provide it for free. Fresh air, clean water, and green grass, Biosphere 2 suggests, will be either a memory or a corporate-produced analogue in the future. Some urban and exurban microenvironments here on Earth already are so noxious that a biospheric envelope would be one kind of rational intervention to counter their toxic effects, and Space Biospheres Ventures is well aware of such commercial applications for its potential technological spin-offs. At some of its biospherics conferences, for example, SBV has featured the work of environmental engineers who have adapted bioregenerative systems into domestic air-handling systems, creating one possible version of "Nature's answer to Earth's environmental pollution problems."[20] Architectural designers have sketched designs for modularized "Bio Homes" to create closed water, air, and sewage recycling systems inside single-family houses to improve air quality, enhance water quality, and produce edible biomass.[21] Of course, admission here would come at a considerable cash price, and not just anyone will be admitted. Likewise, there are other settings—in the polar regions, amid vast deserts, under the sea, or ultimately in extraterrestrial space colonies—where such biospheric systems also might someday be in demand for entire scientific or corporate towns. But, there again, the plan is to produce and deliver at carefully controlled rates in new biosphered sites certain ecological benefits—probably corporate employees producing special high-value goods and services—that one ordinarily would see as a fundamental right of Nature in Biosphere 1.

This vision of corporate ecology, then, reifies and commodifies environments into pay-as-you-go experiences in which ecological benefits can be captured and contained within space frames and under glass. Access into these comparatively simple, but nonetheless still environmentally wholesome, settings can then, in turn, be vended to those with the inclinations and resources to reside in such climate-controlled spaces.

Ecology was the rage in 1989 and 1990 as Biosphere 2 was being readied for its 1991 launching. The entire world prepared for Earth Day 20 in 1990, and looked forward to the Rio environmental summit in 1992. Biosphere 2 still trades on this enthusiasm in promoting itself to the public, even after its radical 1994 restructuring. Nonetheless, the ecology of Biosphere 2 is not that envisioned by the Sierra Club, Earth First!, The Nature Conservancy, or Greenpeace. Instead, it provides a disciplinary space to invent a new science, or "biospherics," dedicated to engineering artificial simulations of terrestrial ecologies in extraterrestrial, nonterrestrial, or harsh terrestrial settings where there are new "possibilities for creating new spheres of life as well as preserving and enhancing the potentiality of the biosphere of the Earth."[22] Biosphere 2, therefore, must not be mistaken for a fundamentalist preservationist enterprise dedicated to restoring some inner balance to life on Earth as it might have been prior to, or would be apart from, the workings of contemporary transnational capitalism. As the multimedia slide show in Biosphere 2's orientation center suggests, these terraforming agendas are the project's bottom line: "The ultimate test of applied ecology—start with nothing and build an ecosystem." Yet, as the slide show admits, starting with nothing really means assembling "a collage of things in Nature" to fabricate something for humanity "to shoot for in space."

Planet Earth is not being reproduced in Biosphere 2; it is instead being resynthesized into something like, but also into many more things quite different than, Planet Earth.[23] Instead, Biosphere 2's ecosystemic modeling is a technology for colonizing alien terrestrial and extraterrestrial realms with scientifically simulated total Earth environments. As some of Biosphere 2's consulting designers see it,

> biospherics, the integrative science of the life sciences as astronautics is the integrative science of the physical sciences, has implications far

beyond the profound possibilities opened up for the understanding of Biosphere I (the biosphere which *Homo sapiens* now, in 1989, inhabit on planet Earth). Namely, biospherics opens up, together with astronautics, the ecotechnical possibilities, even the historic imperative, to expand Earth life into the solar system and beyond that to the stars and then in time's good opportunity to the galaxies, perhaps in association with biospheres from other origins.[24]

These agendas might spin off technologies to control pollution, mitigate hazardous wastes, and rationalize garbage management here on Earth. If they do, then some environmental good will be realized from Biosphere 2. However, at Space Biospheres Ventures' bottom line, biospherics really has much bigger fish to fry out in the solar system and deep space inasmuch as "Biosphere II will provide the first model and the data for its improvements for the fundamental module that will allow the successful building and operation of the Mars settlement."[25]

Vladimir Vernadsky's geochemical reinterpretation of terrestrial biophysics provides the essential insight anchoring the Biosphere 2 project: "the biosphere may, by reason of its essential nature, be regarded as a part of the Earth's crust that is provided with the power to transform the radiations from the cosmos into active terrestrial energy: electrical, chemical, mechanical, and thermal."[26] If it is an energy-collecting, storing, and transforming mechanism, then technologies, such as Biosphere 2's visions of ecotechnics, can be invented to concentrate and apply these forces more efficiently. To Vernadsky, the biota of Nature are not significant simply because of their biological qualities. Instead, he sees them as geochemical phenomena, transforming the planet as geological/chemical/atmospheric forces simply by existing. Thus, Nature and its ecologies in the conceptualization of biospherics are approached as geochemical mechanisms, even though they are organic and vital, to discover their inorganic mechanical properties as ecosystem servers.

Biospherics has been touted openly by Space Biospheres Ventures as humanity's instrument for attaining its true extraterrestrial destiny. As Konstantin Tsiolkovsky foresaw in the nineteenth century, it would be impossible to live on rockets without a self-sustainable supply of biological and physical resources.[27] Russian biomedical engineers and space scientists launched a series of experiments in 1961 to simulate Earth-like environments in microsystems for human habitation during space travel, which evolved into the Bios-3 experiments in the early 1980s.[28] As one higher refinement of Vernadsky's noosphere, or sphere of consciousness and intelligence, of Earth, biospherics provides a path to realize "those developments, both active and contemplative, which assist life in realizing its destiny in the *cosmosphere*, or realm of universal history. . . . part of human potential is to serve as steward to the biosphere here on Earth, and to assist its spread and evolution through space."[29] The copy of a Nature for which there is no original, then, provides a model technology for emulating one kind of terrestrial habitat as extraterrestrial habitations. Rather than serving as an aid to the stewards of the biosphere on Earth, biospherics picks and pulls bits and pieces from Earth's many diverse biomes into its new synthetic simulations of terrestrial biophysics, reducing them to nothing but bioregenerative life support systems for colonizing Terra's biologies as exobiologies on the moon, Mars, various asteroids, or other cosmospheric sites beyond the solar system. This project really gives a new meaning to the notion of "Earth First!" in ecological discourses as the Biospherians of Biosphere 2 saw themselves as Biosphere 1's best opportunity to "birth offspring that can escape to other stars."[30]

Biosphere 2 is not a faithful replication of the earth, or Biosphere 1. It pretends to take The Nature Conservancy to its logical limit, but the facility is instead an engineer's simulation of the denatured Nature surrounding contemporary late-capitalist technoformations, whose forces, in turn, manufacture the denaturing trends in these natural environments for our coevolutionary conditions in the 1990s. A designer Earth, Biosphere 2 is minus millions of species from Biosphere 1 as its emulations can include only those life-forms deemed essential to the reproduction of human life in an artificially encapsulated sphere of material and energy flow. Instead of all plant and animal life, one gets only those species specially deployed to operate as bioengines in the energy conversion apparatuses of Biosphere 2. In this regard, Biosphere 2's architectural complex closely emulates the cyborg planet being constructed now by transnational capitalism in Biosphere 1. Real biodiversity and true wilderness increasingly are lost on the earth in the shuffle of ecology-constituting corporate biomes ecoengineered around an exobiological intention—accumulating more and more capital. Although fragments of Nature are shackled into Biosphere 2 as slave servomechanisms, its basic ecology is essentially cybermechanistic, simulating the increasingly denatured Nature of Earth inside an ecological formation in which humans, computers, mechanisms, and biomasses become one interdependent, coevolutionary energy generation and conversion circuit.

The conceit of Biosphere 2 is that it is an exact copy of raw Nature from Biosphere 1. Ironically, it is, but not in the ways that its designers believe; that is, in perfecting what they see as the pristine logics of Earth's ecology, the ecoengineers of Biosphere 2 trace the denatured Nature of transnational corporate capitalism into its basic operational parameters. Nothing is just existing here as such—as one might see or expect in Nature when it was Biosphere 0. Peter Warshall, a consulting ecologist involved in designing the synthetic balance of Biosphere 2, admits that the overall biodiversity represented in the experiment is only about one-tenth that of any typical ecosystem in Biosphere 1.[31] And, for the insect population, it is only about one-hundredth of what prevails in Nature.[32] There is no wilderness, no arctic, no emptiness, and no fallow in Biosphere 2; they are not functional to this articulation of Spaceship Earth. Seeing Earth as a spaceship is the excuse to terraform Earth to fit the terraforming designs not of creation, but of capital. And, capital's instrumental rationality seeks maximum output for minimum input, creating a Denature, whereas Nature only occasionally works this way. The sublime irrationality, excess, and absurdity of Nature are lost in the biospheric engineer's plans for mobilizing refunctioned plants and animals as subsystems in an ecotechnics equation.

Ecology as Simulation: A Hyperreal Earth

The mechanical ecologies of industrial Earth can be assessed in terms of their fuel, material, and waste product metabolisms. For so many horsepower, gigawatts, millions of gallons, or tons of output, one must input so many millions of barrels of oil, tons of coal, feet of pipe, pounds of chemicals and thereby throughput so many units of mass and energy in making the conversions. These industrial ecologies are still only dimly understood, as the Worldwatch Institute illustrates in its work, but they are the inspiration for biospherics reinterpreting Nature itself as a cybernetic mechanecology. If Nature conforms to these assumptions, then it becomes possible for biospherics to ecoengineer an artificial emulation in autoregenerative units. One might emulate a prehistoric or preindustrial ecology too, but why would anyone want to? Life as we know it now for the billions of Earth's human inhabitants would not and could be lived there without the ultraintensive exploitation of Nature by the economies of transnational corporate capitalist society. This technospheric Denature is bounded space now policed by the nation-state and corporate capital as "the environment," and it is this space of surveillance, management, and production that is the biosphere that Biosphere 2 must imitate. Even though they will not state it in these terms, Nature Conservancy chapters already admit that capital is denuding the planet of any biotic species and microecosystems that its market exchanges can not colonize for profit. Thus, to produce beef, corn, wheat, pork, rice, chicken, rye, mutton, cassava, fish, or oats, all other plants and animals in the spaces needed for these foodstock systems are declared to be weeds or pests to be eradicated in the name of "sustainable development" of these food system feedstocks. Such synthetic environments, in turn, constitute ironically "the natural ecologies" that most of today's environmentalists somewhat vainly struggle to preserve or restore.

By beginning at ground zero, however, Biosphere 2's designers aspired to soup up the biosphere's prevailing planetary operating conditions. One major difference, for example, "will be that the ocean and landmasses will not be in a 70:30 surface proportion, but rather an 8:92 proportion. However, the ocean in Biosphere 2 will operate at least at 10 times the average productivity of Biosphere 1's oceans while the land will operate at about four times the average productivity of comparable tropical terrestrial ecosystems in Biosphere 1."[33] Therefore, the proportion of all live biomass to overall carbon dioxide in Biosphere 2 also was to have been six thousand times greater than Biosphere 1. This fact will "lead to much more rapid carbon dioxide cycling by the system—from a period of about eight years in Biosphere I to half a day in Biosphere II."[34] Here is the apotheosis of sustainable development ideology concretized as an environmental generator: this mechanecology is engineered to produce a discrete planetoidal space complete with its own atmospheric, biospheric, and aquaspheric zones of habitability at indefinitely sustainable levels of operation to serve human beings. What is being proposed for conquering space is, in fact, echoing the means being used now for re-colonizing any place on Earth. Wildness for the sake of wildness in Bio-sphere 1 is an impossible contradiction in Biosphere 2.

Beyond its efforts to cite the geophysical appearance of the earth in its logos, Biosphere 2 is a hyperrealization of the planet's biophysical processes. The varied species chosen to occupy Biosphere 2's life zones all were matched to coexist in this setting under artificial conditions with all of the experiment's other biota. Engineered environmental pro-ductivity rather than autochthonous evolutionary happenstance rests at the core of Biosphere 2's world under glass. In ways that Paolo Soleri's arcologies do not even contemplate, as the next chapter shows, Bio-sphere 2 is truly a new arcology, fusing architecture, engineering, agri-culture, and ecology into a single closed environmental system.

The apparent naturalness of the biomes mystifies its hyperreal prop-erties with elements of air, water, soil, plants, and animals, but, at the same time, most of these operators came through Tucson International Airport as jumbo jet freight for service on this technological ark. The ocean, for example, features artificially generated waves and industrially scrubbed waters to modulate algae populations. It reproduces only the high biomass, high light shallow ocean waters near coral reefs and coastal lagoons; it ignores low light, low biomass deep water and mid-ocean marine regions, it omits big predators and marine mammals, and it overlooks the arctic marine environments entirely. Edible crabs, mus-sels, clams, and lobsters are combined with a hyperreal coral reef, fab-ricated out of life-forms from Caribbean and Gulf of Mexico waters. Pump-driven tides also lap into a marsh estuary modeled on Florida's Everglades where black and white mangroves mediate water flows be-tween the ocean and a freshwater pond.

Meeting model specifications also dominates the environments of the other biomes. The tropical rain forest is modeled after Amazonia, and it combines species of animals and plants taken from all over the Amazon and Orinoco rain forests. The savanna biome brings together grasses, shrubs, and trees from Africa, South America, and Australia, and the desert biome integrates flora from Namibia, Baja California, Chile, and southern Arabia. Biosphere 2 also features birds, insects, fish, amphibians, bats, and reptiles taken from all over the planet to fill various ecological niches in this designer collage of plants, soils, and waters. Finally, the intensive agriculture biome mustered together a fish-rice-azolla aquaculture zone, a small goat/pig/chicken ranch, a tiny herb garden, a miniaturized fruit orchard, a legume and tuber plot, and a diverse grain farm in eighteen fields. Rotating through three crops a year, this zone supposedly mimics subtropical regions with high humidity and temperate ranges from 65 degrees Fahrenheit winter lows to 85 degrees Fahrenheit summer highs. There is no original analogue to this biome—in terms of depth or diversity of some really existing agricultural region—anywhere in the world. Instead, this biome was engineered around a mix of foods representing the flow of products to corporate supermarkets made available by global food commerce; hence, it emulated through intensive on-site production what the average suburban consumer can feed upon after making extensive car trips to the store in an American or European city.

Forces in Nature played out through vulcanism, weather, tidal flows, sunlight, rain, river currents, wind, fire, and topographic variations also are mechanically simulated in Biosphere 2. Pumps, fans, and pipes generate hyperreal weather patterns, while metal girders, spray concrete, and steel plate re-create rock formations, hillsides, and bedrock foundations. A glazed space frame mimics gravity and atmosphere to provide breathable air, composters generate tillable soil, and gas-fired turbines produce hot and cold water, electrical power, and climate control. This replication of the biosphere through a complex technospheric apparatus, in turn, completes the manufactured viability of all the other ecosystemic models coevolving as a simulation of Earth. Designing, constructing, and operating a fully enclosed, airtight ecological system on three acres of land is a remarkable engineering achievement, but it is a dubious environmental milestone. Space Biospheres Ventures suggests that all of its expertise now might be put to use by refining human environmental impact on the earth, but the alleged solutions tested here would efface Nature's random eventuations of life's varied diversity with instrumentally rational, replicant ecosystems pitched exclusively to serve their human hosts.

Biosphere 2 incarnates the effort to reduce all of Biosphere 1 to a hyperreal technosphere as humanity's bioregenerative support system. It turns anthropocentrism into rational arcology, fusing human needs into architecturalized closed ecologies as it erases the life-forms of real Nature that might endanger humans, disrupt food production, limit ecosystemic viability, or serve no obvious function. Biosphere 2's biospheric technologies have no place for tiger sharks, grizzly bears, houseflies, army ants, Bermuda grass, or Russian thistles. Biosphere 2 is a high-tech designer planet, drawn to omit the pests and weeds its inventors have decreed to be dispensable. With these moves, it belies its aspirations to emulate authentically Biosphere 1 in concocting a technoscientific simulation meant for colonizing either uninhabitable terrestrial spaces (and thus further stress its habitable spaces to build this ecologizing empire) or the uninhabited reaches of extraterrestrial spaces (as an environmentally overstretched Earth requires off-planet energy and material inputs and/or resettlement sites) with human populations expanding beyond the earth's current carrying capacities.

At the bottom line, Biosphere 2 is an elaborate technology for materializing anthropocentric/anthropogenic change on an ecosystemic scale. It is not devoted to Nature or even to naturalism, because Space Biospheres Ventures wants to show how "underproductive" and "inefficient" Nature lets its systems operate. Just as Du Pont and nineteenth-century capitalism brought humans better living through chemistry in the twentieth century, SBV and twentieth-century capitalism with its partners in the Global Systems Initiative promise humanity better living through the biospherics of planetary engineering in the twenty-first century. As Dr. Michael Crow, vice provost and head of the Global Systems Initiative at Columbia University, notes with regard to Biosphere 2, "one hundred fifty years from now, there will be planetary engineering departments at major institutions like Columbia."[35] Thanks to tests at sites like Biosphere 2, one school of planetary engineering "will try to maintain the Earth as its natural self," and another will say "we're the masters, we can control all of this. You want more CO_2 in the atmosphere, we'll give you that. You want the oceans five degrees warmer, we'll do that. You want the plates to shift this way or that—no problem."[36] With Biosphere 2, worldwatching systems of environmentality can be translated into planetary engineering projects to extract maximum advantage from Nature as Denature for future resource managerialists. Indeed, Biosphere 2 could become a launchpad for many new Global Systems Initiatives to initiate a global systematization in green governmentality maneuvers to control the conduct of conduct by assuming the guise of such systems for terraforming environmentality.

6

Green Consumerism:
Ecology and the Ruse of Recycling

In the material culture made possible by today's global capitalist economy, as chapters 4 and 5 show, the personal sphere of the everyday life world is thoroughly political. The ways that material wealth is produced, distributed, and consumed all represent the outcome of innumerable depoliticized technical decisions made by product designers, industrial engineers, corporate managers, public administrators, and marketing executives, all striving to attain the most rational solutions to their respective technical challenges for the economy's abstract machines. And, in exchange for a constantly increasing level of material comfort and economic security, virtually every client and customer of this global capitalist economy accepts the outcome or impact of these decisions with little or no protest. Larger cultural trends, then, in global economic and social rationalization tend to proceed apace without any popular representation in the processes of their instrumental or substantive determination.

The scope of these powers over everyday life is extensive. Such forces determine who gets what, when, and how in the most immediate material sense. Yet, the political dimension continues to be mystified in discourses of technical expertise, economic imperatives, or social necessity, which all allegedly exert their essential effects of everyday life beyond the realm of ordinary politics. Most mainstream analyses of politics ignore such subtle contradictions, but, since the 1960s, small social movements based on an ecological radicalism have contested these cultural dynamics. By exposing the many mystifications of environmental balance with issues of technology and industry, ecological radicals have struggled to win popular representation within, or even popular control over, these processes of rationalization. But the political terrain in this battle is always shifting.

In today's overheated political discourses, the spin placed by the mass media on concern for the environment has shifted profoundly from Earth Day 1970 to Earth Day 1990. In the 1960s and early 1970s, any serious personal interest in the environment often was seen as the definitive mark of radical extremism.[1] To deter the attacks of environmentalists, big business frequently argued that growth was good, that any legislation aimed at limiting pollution meant cutting jobs, and that ecologists were crackpot limousine liberals willing to put the existence of snail darters before modern humanity's material progress.[2] However, by 1990, many assumptions in the popular discourse about nature, ecology, and environmentalism had changed significantly. Holdouts from the old school still follow the old ways when, for example, they pit the material welfare of loggers against the survival of the spotted owl, or tout the necessity of real-estate development over protecting of coastal wetlands in this or that corner of the nation. But, on the whole, Earth Day 1990 saw "environmentalism" become a much more legitimate—or even mainstream— public good. Many major corporations now feel moved to proclaim how much "every day is Earth Day" in their shop, what a meaningful ecological relationship they have with nature, or why their manufactures are produced with constant care for the planet's biosphere.[3]

These claims are still largely false. Yet, this elaborate change in ideological emphasis by corporate political discourse is very significant. What has changed in the social imagination of ecology and environmentalism that makes these astounding rhetorical ploys not only possible, but also apparently somewhat convincing? As this chapter will indicate, one vitally important change can be traced back to the domestication of some types of ecological radicalism as fairly tame forms of environmental reformism, or "green consumerism." The rhetorical commitments and political practices of green consumerism are important because they have provided many people with the most accessible and widely distributed discussions of the environmental crisis through a large number of ecological handbooks and self-help guides that urge consumers "to go green." In turn, then, this chapter asks how the popular understanding of the ecological crisis has been altered or reconstructed by green consumerism such that these ideological maneuvers have acquired popular acceptance. Answers to such puzzling questions can be found by critically reconsidering green consumerism's interpretations of what are some of the late-twentieth-century's most trusted articles of faith, namely, ecology, environmentalism, and recycling.

Shifting Rhetorics: From Production to Consumption

In the 1950s, 1960s, and 1970s, corporate managers and government bureaucrats often tried to crush ecological protests, figuring that such direct strategies of political intervention rhetorically might roll back the symbolic assaults of troublesome "tree-hugging" nature lovers by continuing to convince the general public that belching smokestacks still were signs of material progress and not the sigmata of industrial pollution.[4] During the early 1970s, however, President Nixon and Congress rapidly reacted to mounting popular pressure for environmental reforms with a series of major legislative initiatives and passed the Clean Air Act, the Clean Water Act, the Endangered Species Act, and the Resource Conservation and Recovery Act. They also authorized the establishment of the Environmental Protection Agency (EPA).[5]

By many objective measures, these laws did begin to have a discernible impact on pollution levels in many areas. Similarly, the elaborate oversight mechanisms of the EPA began to create at least the appearance of corporate compliance with government environmental codes. And, finally, the flight of factories from the United States to the Third World in search of cheaper labor and more lax pollution laws often made production less of a worry. Therefore, environmentalist organizations found that the image of the ecological enemy often had to be rhetorically expanded and intensified. In an ideological turnaround, which was ironically aided and abetted by some groups of ecological activists, it became clear that some of the worst environmental offenders no longer are simply dirty polluting factories or hungry lumber mills, but rather allegedly are individual consumers. With the regulatory fig leaf provided by EPA regulations and federal environmental law, the rhetoric of ecological responsibility slowly shifted from a vernacular of "Big business is dirty business" to dialects of "Factories don't pollute. People do."

To contain the emergence of new, and perhaps more radical, measures of ecological transformation, the corporate-run circuits of mass consumption refunctioned the rhetoric of that faction in the ecology movement whose alternative social agendas favored green consumerism as a means of reducing material consumption into a new special subsystem of mass consumption. In these rhetorics of reform, as they ironically are used by both environmentalists and corporate public relations, major corporations—such as Exxon, General Foods, and Phelps Dodge—are not responsible for pollution. Rather, each individual consumer or family household is now the key decisive ecological subject, whose everyday economic activities are either a blow for environmental destruction or a greener Earth. Amazingly, the one wing of the environmental movement that once was opposed to producing more material goods, accepting guided mass consumption, and embracing subtly engineered preferences has gained greater acceptance and visibility in the 1990s by presenting itself, in part, as a new set of material choices, a fresh guide for mass consumption, and a revolutionary reengineered set of "green" consumer preferences.

Throughout the go-go 1980s, this alternative vision for solving the ecological crisis largely was ignored; yet, in the aftermath of the 1989 Exxon Valdez tanker accident, the mass media and oil industry spin doctors seized on these alternative lifestyle agendas to construct a narrative of collective guilt that could explain why that tragic shipwreck and oil spill "had to happen." Once launched, these counterrevolutionary ecological rhetorics also gained tremendous momentum from 1989 to 1991 with a rush of new mass-marketed handbooks for ecological living, whose authors and publishing companies had geared them up into production to capture the growing green market arising from the increasing levels of hoopla for Earth Day 1990. Clearly, various guidebooks for clever consumption have been produced for decades to help consumers make their buying decisions. As part of the 1990 Earth Day mobilization, even the infamous Heloise of syndicated newspaper feature fame issued her *Hints for a Healthy Planet* to help her millions of now green-leaning readers cope with the demands of environmental housekeeping.[6] However, a whole new crowd of even more elaborate consumer guidebooks joined Heloise in 1989, 1990, and 1991 to lead consumers to a green future as their authors urged readers to forsake a long march toward the institutions in favor of a long shopping trip through the malls to revolutionize modern society. Hence, in every one of the hundreds of Waldenbooks or B. Dalton's bookstores across America, thousands of concerned consumers could ring up another purchase to learn how to save the planet.

Looking at Guidebooks for Green Consumption

Saving the earth and preserving the environment are extremely complex challenges, and there are no easy or simple solutions for today's ecological problems. Yet, in complete fulfillment of the fallacy of generalization—namely, if every X did Y, then Z would certainly follow—the Berkeley, California-based Earth Works Group confidently markets a whole series of self-help manuals based upon "50 simple things" that everyone can do to "save the Earth." The first book, *50 Simple Things You Can Do to Save the Earth*, alleges that it "empowers the individual to get up and *do something* about global environmental problems."[7] "Most of the 50 Things," the reader is told, "are unbelievably easy. They are the kind of things you would do anyway to save money—if you knew how much you could save."[8] The book claims that, rather than allowing negative media coverage of the environmental crisis to drive one to despair, each consumer can do some "unbelievably easy" things to conserve cash as well as to solve "intractable environmental problems."[9] If every consumer bought this one book and followed its directions, ecological salvation would surely follow.

How does the Earth Works Group's solution hope to work? First, by acknowledging the real powerlessness of consumers; and, second, by whittling away at major supply-side irrationalities through urging consumers to make slightly more frugal and marginally more rational choices about obtaining the material wherewithal needed for their day-to-day survival. Instead of thinking about how to reconstitute the entire mode of modern production politically in one systematic transformation to meet ecological constraints, the book, like most tracts of green consumerist agitation, bases its call for action on nonpolitical, nonsocial, noninstitutional solutions to environmental problems "that cumulate from the seemingly inconsequential actions of millions of individuals.

My trash, your use of inefficient cars, someone else's water use—all make the planet less livable for the children of today and tomorrow."[10] Consequently, the corporate institutions that produce goods wrapped in this trash, that restructure cities to require travel in their inefficient cars, and that build appliances, homes, and cityscapes based on wasting water are automatically excused almost from the outset, except inasmuch as individuals might effect change by choosing to use less of their products or deciding to purchase alternative merchandise.

The logic of these corporate institutions' resistance, then, just like the logic of the average consumer's initial compliance, is centered on the still largely passive sphere of *consumption* rather than on the vital sites of *production*. The ecological battle lines are drawn at the gas pump or in the supermarket aisles not at the factory gates or in the corporate board-rooms. The Earth Works strategy asserts:

> Few of us can do anything to keep million-barrel oil tankers on course through pristine waters. All of us can do something, every day, to insure that fewer such tankers are needed. None of us can close the hole in the ozone layer above Antarctica. All of us can help prevent its spread to populated areas by reducing our use of chlorofluorocarbons (CFCs).[11]

This characterization of environmental conflicts rightly notes, on one level, that using less gasoline and underarm sprays in the United States might well lessen oil tanker traffic and reduce the hole in the ozone layer. But, on another level, it wrongly suggests that people cannot really expect to use collective political means to keep tanker accidents from happening or to totally eliminate CFCs. The whole ecological crisis ultimately is reinterpreted as a series of bad household and/or personal buying decisions: "as much as we are the root of the problem, we are also genesis of its solution."[12]

The aggregate effects of the ecological crisis, therefore, can only be framed in terms of the accumulated collective impact of consumers' choices. The key dimensions of the crisis, according to Earth Works, are the greenhouse effect, air pollution, ozone depletion, hazardous waste, acid rain, vanishing wildlife, groundwater pollution, garbage, and saving energy. In turn, allegedly reckless individual consumption, which now supposedly causes all of these problems, can, at the same time, solve them by individuals shifting consciously to patterns of reasonable individual conservation. Conservation, ecological sustainability, frugality "can be accomplished by simple, cost-effective measures that require little change in lifestyle."[13] Here is the other major flaw in the Earth Works approach. In fact, more reasonable patterns of individual consumption were once quite common, but corporate imperatives to stimulate mass consumption of mass-produced goods have overridden these traditional restraints with today's throwaway lifestyles. Corporations have spent decades developing complex, cost-effective techniques that have required massive changes in each consumer's lifestyle, which is based on wasting large amounts of energy and resources by purchasing corporate-provided commodities.

After defining the ecological crisis as essentially a struggle over how elites go about shaping tastes in the everyday life world of material consumption, the Earth Works Group outlines a three-stage strategy for reshaping these patterns of taste, which are organized around three different levels of relative effort required from consumers to make them succeed. First, there are really only twenty-eight "simple things" to begin doing, such as stopping junk mail, keeping automobile tires fully inflated, giving up styrofoam cups, refusing to buy ivory products, or using energy-saving light bulbs. Second, there are twelve things that "take some effort," such as recycling glass, using cloth diapers, planting trees, or car pooling. And, third, there are eight things "for the committed" that require bigger, and actually more necessary, changes, such as driving less, eating less meat, replanting one's house lot in xeriscaping, and getting involved in community associations working for ecological change. Most of these recommended changes, however, merely involve resurrecting an ethic of frugality, thrift, or common sense. This ideological line is perhaps most obvious in the Earth Works Group's exclusively focused recycling guide, or *The Recycler's Handbook: Simple Things You Can Do.*[14] Those far-reaching changes that push beyond the limited agenda of recycling actually would appeal to very few people because they would require more than a "little change" in lifestyle.

Indeed, the Earth Works Group implicitly recognizes these limitations since it also has produced a follow-up book, *The Next Step: 50*

More Things You Can Do to Save the Earth, that suggests "snipping six-pack rings may be a start, but it's not the solution. . . . It's time to reach out to the community."[15] This handbook does begin to ask some political questions, but its style of politics is posed almost entirely in the tame dialects of Naderite public interest insurgency.[16] The "next step" of "fifty more things to save the earth" simply takes green consumerism down already familiar tracks, such as using affinity-group charge cards, pushing for local curbside recycling programs, starting a ride-sharing system, buying only recycled goods, urging retailers not to sell ozone-damaging goods, or starting a municipal yard composting program. It hints at promising political action, but the political activities being advanced mainly are directed at motivating more people to start doing the first fifty simple things to save the earth. This weak reformist strategy even is affirmed in the Earth Works Group's appeal to more radical youth audiences, *The Student Environmental Action Guide: 25 Simple Things We Can Do.*[17] To paraphrase Marx, Earth Works environmentalism fails inasmuch as it has only thus far been recycling one tame interpretation of the world, when the real point is to discover how to change it.

The real intellectual limits of the Earth Works Group's tame interpretations of environmental transformation become more obvious in one of its latest works, *50 Simple Things Your Business Can Do to Save the Earth.*[18] Rather than directly attacking the obvious ecological irrationalities in most businesses' production processes, this manual "recognizes the realities of business" by claiming that its Earth Works approach "can yield dividends in this fiscal year—in cost savings, lower taxes, improved company image, and in increased employee satisfaction and productivity. This is a textbook case of 'doing well by doing good.'"[19] Each business is treated mainly as "a superconsumer" that can, like other individual consumers or private households, also contribute to ecological change by doing the same "simple things," such as reorganizing the office coffee pool to use ceramic mugs, recycling office paper, buying green cleaning supplies, changing to low-energy light fixtures, fixing company toilets to use less water, composting landscape by-products, or remodeling the office with plants, nonrain-forest wood products, and solar climate control.

Unfortunately, the logic of resistance behind these changes is totally defensive. The Earth Works Group accepts the modes of industrial production as they operate now, but urges that employees engage in their own environmental policing to avoid running afoul of the prevailing legal, bureaucratic, and public relations problems that regularly befall many companies. Green is good because it saves money, it is good public relations, and, finally, of course, it is good for the environment to boot. Rather than pushing waste elimination, the Earth Works Group stands for waste reduction. Instead advocating total economic transformation, it accepts weak bureaucratic regulation of present-day polluting processes. Unable to support the total reconstitution of today's productive forces, it advocates piecemeal reforms to lessen, but never end, their most environmentally destructive activities.

The Earth Works Group also fails to identify the key potentialities of workers and management in modern businesses for realizing ecological changes. For example, Earth Works notes that "if you work in an office, a workshop, a factory, you are the backbone of your company. You and co-workers can use your collective influence to mold policy decisions."[20] This claim sounds, at first, quite impressive, but with this allegedly immense collective influence, it directs workers to debate making decisions about essentially insignificant choices: "Should you throw out that piece of paper . . . or recycle it? Is it too much trouble to wash out a mug so you don't have to use a disposable cup? Should you leave a light or copier running . . . or turn it off?"[21] If the backbone of corporate America is misdirected into agonizing over policy decisions like these, then critical ecological choices about what to produce, how to produce it, when to market it, and where to distribute it will all be left to those managers in high positions who know "it's not possible to turn well-honed products and processes topsy-turvy to protect the environment and still function as a business."[22] The Earth Works Group, then, winks at prevailing practices of antiecological management, privileging the passive acceptance of corporate managers' expertise and the legitimacy of not troubling dedicated executives as they discharge their tough decision-making tasks. Instead, it pushes ineffectual window-dressing practices on ordinary employees to green marginal aspects of their firm's office ecologies or their company's public image. The fact that everyone in a company uses an ecologically correct ceramic mug and recycles office memos does not lessen the environmental destruction that this same firm might be spreading by building gas-guzzlers, selling CFCs, mowing down rain forests, or manufacturing plastic playthings.

Along with the Earth Works Group and its widely distributed books, Jeremy Rifkin's edited volume *The Green Lifestyle Handbook: 1001 Ways You Can Heal the Earth* has been a major hit in the nation's bookstores.[23] In spite of its frothy political pitch, it too pushes strategies for household revolutionization to propagate the virtues of an "ecological lifestyle." Combining the separate household, student, and business product lines of three books from the Earth Works Group, Rifkin's one anthology invites its readers to join the green revolution by embracing an ethic of voluntary simplicity or purposeful frugality. The opening section of the handbook, or "back to basics," sketches out the virtues of curbing home energy use, using green household cleaning products, shopping with an ecological attitude at the supermarket, and reorganizing office waste streams for recycling. The second section, on "lifestyles," bids the reader to acknowledge the importance of eating down low on the food chain, returning to simple idleness for leisure, and investing in a socially responsible fashion. The third section, on "cultivating solutions," affirms the ecological wisdom of personal gardening, tree planting, low-impact agriculture, and preserving genetic diversity.

This consumeristic treatment of ecology, of course, covers the same individually centered ground as the Earth Works Group's fifty simple things "at home" and "in business" that one can do to save the earth. Picking up on the public-interest citizen lobbying initiatives of the Earth Works Group's *The Next Step* book, the last section of Rifkin's handbook underscores the critical importance of "getting organized" to realize a local, grassroots "environmental democracy." By targeting corporate polluters, utilizing boycotts of antiecological products, lobbying elected representatives, and becoming informed about environmental litigation, Rifkin's handbook suggests that an "organized citizenry" that follows these obvious techniques of ordinary citizenship can win "a place at the bargaining table to ensure that local corporations stop local poisoning and preserve the planet's limited natural resources."[24] The fact that such places are rarely won, that their occupants also typically have little voice, or that local poisoning, if stopped, is only moved elsewhere is not really discussed. Still, Rifkin invites all of his readers to push through these sorts of negotiations, because public-interest environmental advocacy should gain "more influence over the activities of companies that threaten the environment. Also, we must organize to insure that local, state, and federal legislators pass the necessary laws to help save the planet."[25] Of course, this call to activism is not really all that new; this set of tactics has structured the agenda of mainstream environmentalism since the mid-1960s.[26] Sometimes it works successfully for a bit on the margins, but mostly it falls short by a very wide mark.

Another best-seller, Jeffrey Hollander's *How to Make the World a Better Place: A Guide to Doing Good*, also promises "over 100 quick-and-easy actions," as if ecological revolution were an instant cake mix, to show "how you can effect positive social change."[27] This book can be read, as the author's foreword suggests, "in any order you choose" so that one will not "get bogged down by all the introductory facts" and anyone can "always skip ahead to the action."[28] Arguing that it is not yet another journalistic description or just one more academic analysis of the world's ecological crisis, this handbook offers ways to link everyone's "sincere and noble desire to help and the concrete, effective actions necessary to effect change."[29] Like Rifkin and the Earth Works Group, Hollander asserts that "the shape of the future is in our hands. It is our responsibility, for it can be no one else's. The world won't be destroyed tomorrow, but it can be made better today."[30]

How these hitherto unattained radical advances toward global ecological harmony will be wrought "is designed to generate the greatest impact in the least amount of time" and, miraculously, "results are guaranteed without marching on Washington, quitting your job, or giving away your life savings."[31] On one level, Hollander's endorsement of Naderite public-interest lobbying tactics ensures that his readers will learn "how to help build low-income housing, use recycled products, contribute to a food bank, invest money in a socially responsible manner, free prisoners of conscience, pass legislation through the U.S. Congress, and encourage world peace."[32] And, even more fortunately for today's harried average consumer, Hollander claims, like an ad for taped foreign language lessons or some new tummy-reducer gizmo, that "only a few minutes are needed for many of the actions that will result in positive social change."³³ On another level, Hollander makes the ultimate radical claim for today's socially concerned, but fundamentally passive, consumer: the ecological revolution really can be made essentially by doing nothing more than ordinary everyday things. In other words, you can learn how "to make the world a better place as you wheel your cart down the aisle of a supermarket, travel on business or pleasure, select an insurance policy, open a bank account, prepare dinner, relax around the house, and even as you soap up in the shower."³⁴

It may be true that "the actions of those now living will determine the future, and possibly the very survival of the species,"³⁵ but it is, in fact, mostly a mystification. Only the actions of a very small handful of the humans who are now living, namely, those in significant positions of decisive managerial power in business or central executive authority in government, can truly do something to determine the future. Hollander's belief that thousands of his readers, who will replace their light bulbs, water heaters, automobiles, or toilets with ecologically improved alternatives, can decisively affect the survival of the species is pure ideology. It may sell new kinds of toilets, cars, appliances, and light bulbs, but it does not guarantee planetary survival.

Hollander does not stop here. He even asserts that everyone on the planet, not merely the average consumers in affluent societies, is to blame for the ecological crisis. Therefore, he maintains, rightly and wrongly, that "no attempt to protect the environment will be successful in the long run unless ordinary people—the California executive, the Mexican peasant, the Soviet [sic] factory worker, the Chinese farmer—are willing to adjust their life-styles and values. Our wasteful, careless ways must become a thing of the past."³⁶ The wasteful, careless ways of the California executive plainly must be ecologically reconstituted, but the impoverished practices of Mexican peasants and Chinese farmers, short of what many others would see as their presumed contributions to "overpopulation," are probably already at levels of consumption that Hollander happily would ratify as ecologically sustainable if the California executive could only attain and abide by them.

As Hollander asserts, "every aspect of our lives has some environmental impact," and, in some sense, everyone, he claims, "must acknowledge the responsibility we were all given as citizens of the planet and act on the hundreds of opportunities to save our planet that present themselves every day."[37] Nevertheless, the typical consumer does not control the critical aspects of his or her existence in ways that have any major environmental impact. Nor do we all encounter hundreds of opportunities every day to do much to save the planet. The absurd claim that average consumers only need to shop, bicycle, or garden their way to an ecological future merely moves most of the responsibility and much of the blame away from the institutional centers of power whose decisions actually maintain the wasteful, careless ways of material exchange that Hollander would end by having everyone recycle all their soda cans.

Marjorie Lamb takes the demands of mounting a green revolution from within the sphere of everyday life down almost to the bare minimum in her *Two Minutes a Day for a Greener Planet: Quick and Simple Things You Can Do to Save Our Earth*.[38] Her manifesto speaks directly to the harried but still very guilty modern suburbanite: "We are all busy people. Let's face it, we don't have the time or desire to climb smokestacks or confront whaling vessels. But there are lots of things we can do differently every day. Without effort and with very little thought, we can make a difference to our planet Earth."[39] This astounding revelation is precisely what every consumer wants to hear. Like ecological destruction itself, ecological salvation is possible "without effort" and "with very little thought."

Only "two minutes a day" are needed by today's one-minute managers to execute the "quick and simple things" needed to save "our Earth." To fill the bookstores at the mall with yet another cookbook for ecological transformation, Lamb has expanded her original "Two Minute Ecologist" radio spots, first developed for the CBC's *Metro Morning* radio broadcast, into a pocketbook guide to green liberation. And, once again, she stresses the vital importance of recycling aluminum, refusing to buy overpackaged goods, and composting kitchen/household waste. But, interestingly enough, Lamb also honestly remarks that much of her advice is essentially remedial consumer education. Indeed, many, if not most, of the simple hints that she, Hollander, Rifkin, and the Earth Works Group are spelling out were once widely practiced forms of popular common sense. Lamb credits "the Depression generation," or those who grew up prior to 1945, with an ethic of thriftiness that actually approximates many of the virtues she assigns to the coming "Age of the Environment." On the other hand, "the baby boom generation," and now their offspring, have embraced all of the unsustainable habits of mass consumption that corporate capital once encouraged but now recognize are at the root of today's ecological crisis.

In part, Lamb's analysis is true, but it ignores how corporate capital, big government, and professional experts pushed the practices of the throwaway affluent society on consumers after 1945 as a political strategy to sustain economic growth, forestall mass discontent, and empower scientific authority. People did choose to live this way, but their choices were made from a very narrow array of alternatives presented to them as rigidly structured, prepackaged menus of very limited options. And, now ironically, all of these green guides to ecological consumption are moral primers pitched at resurrecting—through their own green but still nicely designed plans of commodified ecological revitalization—the responsible habits of more frugal consumers or autonomous citizens that corporate capital and the mass media have been struggling to destroy for nearly a century.

Green Consumerism as Marketing Demography

Campaigns for individual change, like these mobilizations for green consumerism, rarely change society in any fundamental sense. However, they often do diversify today's already complex consumer markets in new exciting ways by generating fresh psychodemographic niches of need. The Roper organization, for example, recently conducted a series of public opinion surveys that profile the entire American public as members in one of five different behavioral bands when it comes to questions of ecology and recycling. In 1990, 78 percent of all Americans supported "improving the environment."[40] However, only 11 percent were "True-Blue Greens," or those who were regular recyclers, members of environmental groups, and serious supporters of stringent environmental regulations. Another 11 percent were "Greenback Greens," or persons who failed to practice recycling regularly and remained diffident environmental supporters, although they would willingly buy "green" products. A healthy 26 percent were "Sprouts," who recycle regularly but doubt that it really does much good as a purely individual effort. And, the large antienvironmental majority of 52 percent was divided between "Basic Browns" (28 percent) and "Grousers" (24 percent), who either do nothing for the environment or simply make excuses for not recycling and buying green products.[41]

Green consumerism, as these sociologically constructed market segments reveal, is not an insignificant force. Indeed, the allied block of True Blue Greens and Sprouts adds up to 36 percent of the nation's mass markets. This indicator closely parallels other market studies that reveal that about three out of every ten consumers buy products because of green advertising, ecological labeling, or environmentalist endorsements. Therefore, green consumerism, which allegedly began as a campaign to subvert and reduce mass marketing, now ironically assists the definition and expansion of mass marketing by producing new kinds of consumer desire.

To reach nearly a third of all contemporary consumers, no savvy advertiser or manufacturer can ignore the need for retooling existing systems of promotional hooks and lures. For example, green advertising can buy time on a national news program that displays file footage of bird and seal corpses, soaked in Exxon crude and floating in Prince William Sound, to run ads for Conoco's double-hulled tankers that apparently move sea mammals, seals, and birds to jump, leap, and flap in exaltation to the beat of Beethoven's Ninth Symphony and Schiller's "Ode to Joy" as a celebration of the Du Pont Corporation's "ecological concern." But oil continues to be lifted, shipped, and sold to burn in millions of personal automobiles that all are still pumping more carbon dioxide into the atmosphere. Likewise, green interpretations of ordinary manufacturing can rhetorically transform "a woman's selection of a nightgown made of 100 percent cotton into an 'environmentally responsible' purchase of a product made from a 'renewable resource.'"[42] But the cotton most likely still is grown, harvested, and produced using inefficient and dangerous inputs of oil-based pesticides, fertilizers, and fuels that ravage local soils, watercourses, and biotic communities. The rhetorics of modern mass consumption prove capable of even turning an antithetical assault on all of their material premises into yet another psychodemographic space of highly segmented, but nonetheless still fundamentally consumeristic, behavior.

Even big business now is sensitive enough to recognize that a crisis of ecological credibility exists in the green market. Any company can put "ozone-friendly," "ecologically approved," "environmentally sound," or "dolphin-safe" logos, usually printed in green ink, on any product. The percentage of new products that claimed a green pedigree increased from 4.5 percent to 11.4 percent of all new packaged goods from 1989 to 1990. By the same token, the number of advertisements with a green tie-in more than quadrupled.[43] Yet, there is no common legal and scientific definition of "recycled," "biodegradable," or "ecological" available for standard use in labeling or advertising. Hence, many products claim to be all of these good things, while being none of them. Fortunately, however, different advocates of green consumerism are coming to the rescue. Dennis Hayes, the chair of both the 1970 and the 1990 Earth Day celebrations, has organized Green Seal in Palo Alto, California, to issue a "Green Seal" mark of environmental soundness. By allying with Underwriters Laboratories for product testing, and by building coalitions with a few major producers, Green Seal saw itself certifying new products by 1992. A second group, Green Cross of Oakland, California, has agreements with several large retail outlets, and it has already certified nearly four hundred products with its "Green Cross."[44] Although there are considerable differences between the two groups over how to certify a product either in terms of its immediate ecological impact in consumption or over its entire production life cycle, this sort of ecolabeling definitely is winning support.

Green consumerism may culminate in some mix of these conventionalized systems of "safe seals" or "wholesome hallmarks" in the United States to certify the ecological purity of any given product. Indeed, the Blue Angel seal of environmental soundness has been used on thousands of products in Germany for more than a decade, and comparable government-backed programs are beginning in Japan and Canada.[45] Because being an effective, environmentally conscious consumer demands much more than two minutes a day, these different underwriting laboratories of credibility creation already are battling over the right to authoritatively certify whether this or that product is "ecologically correct." This development, of course, simply brings the final pacification of ecological protest by junking tactics of "two minutes a day" and moving to strategies of "two seconds per purchase" in mobilizing the public's understanding of environmental activism. A consumer scans a product, finds the socially accepted seal of approval, and tosses it into the shopping cart with no muss, no fuss, no never mind. The seal suspends the need for environmental consciousness or ecological reasoning; it too already has been expertly engineered by the manufacturer with green certification into the product. Yet, beyond adding the seal, very little, if anything, may have actually changed anywhere in the cycle of producing, distributing, consuming, and disposing of the product. The consumer society still draws in resources from all over the world to the highest wage consumer markets to the detriment of biomes, animals, plants, and other humans to produce new profits through green indulgences for those who can afford to buy them. A broad range of ecologically diverse societies of immediate producers who are environmentally well matched to the peculiarities of their various bioregions does not develop; indeed, its basic plausibility only becomes even more fantastic as "green" goods marked by the appropriate seals of good planet keeping pile up, first, as merchandise and, then, as trash in the consumer society's supermarkets, recycling bins, and landfills.

The amazing popularity of recycling, thanks to Lamb, Hollander, Rifkin, and Earth Works, also actually might be, at the same time, undercutting its legitimacy. The symbolic importance of every household "doing something," especially if it is quick and easy, makes recycling a pervasive practice. Yet, in certain instances, recycling actually may be less efficient or cost-effective than some other resource conservation strategy. Similarly, recycling can fix into place many prevailing packaging or resource use technologies, which may be irrational or inefficient in comparison to alternative techniques, by providing a new, cheap voluntary infrastructure for their retention with recycling networks. Recycling also can create recycled resource gluts of newsprint, glass, plastic, or paper waste that certain communities must pay to collect and store although they may be unable to complete the critical cycles of actual reuse because of saturated markets, a lack of recycling plants, or excessive transportation costs for moving recyclable commodities. Hence, these semirecycled resource stocks actually might have to be landfilled anyway.

Incinerating recycled resources can provide a short-term solution for some combustible products, but this practice also creates air pollution, ash by-products, and new transaction costs. Raising landfill costs can reduce waste by forcing communities and households to conserve their creation of trash, but this option also will only generate more incineration or recycling. Thus, only in recycling does the real source of the problem now perhaps emerge, namely, the manufacturers that continue to excessively package their products, irresponsibly use nonrecyclable materials, irrationally waste energy resources, and unfairly externalize production costs. By shifting the costs of coping with their inefficiency and irrationality to newly motivated recycling households and municipal waste management agencies, producers can continue the exploitative practices of old-wave traditional industrialism as their negative effects are mitigated by new-wave green consumerism. Once again, a core supply-side changelessness is preserved by enveloping it in a demand-side mobilization for marginal change.

Earth Day as the Holiday of Green Consumerism

Making ecological revolution through green marketing and ecological consuming feeds directly into the organization and administration of spectacles like Earth Day to occasionally affirm, or even permanently ritualize, the many diverse practices of green consumerism. Just as mainstream consumer society finds its most complete affirmation in the highly commercialized festival day of Christmas, green consumerism has been woven into the mythologies and rituals of its own extremely commercialized festival day, or Earth Day. Arguably, the Earth Day celebration does serve to promote worthwhile environmental changes and popularize meaningful ecological lessons among mass audiences that might otherwise ignore the concerns of new ecological movements. Yet it also feeds directly into the same destructive logic of consumer festival days like Presidents' Day, Memorial Day, Labor Day, or Columbus Day, as well as Mother's Day, Father's Day, Grandparents' Day, or Secretary's Day, that corporate marketing departments seize on to encourage consumers to make another trip to the mall to show someone that they care enough to convey their caring through more commodities. As a specially valorized day to boost consumption by "showing you care" or "telling someone how much you love them," Earth Day becomes a day to mark how we can save the planet by producing and consuming spectacles about planetary salvation. In turn, major retailers rack up huge sales of can crushers, composters, newspaper bundlers, bicycles, and green guidebooks to fulfill consumer desires to possess the correct icons for observing the day's rituals. Otherwise, like the 1980 celebration, it is totally ignored or marked only by a few nature lovers in total obscurity out in the woods.

Ironically, what began as a festival to call planned mass consumption into question now can survive only if its designers allow it to be packaged by contemporary corporate capitalist society as yet another organized event based on specially planned mass consumption. Wendell Berry's fear that environmentalists are too cautious in their protests as they approach the earth as "nature under glass" is not a problem here. Instead, Earth Day commodifies nature and concern for nature as still another set of carefully coded products to circulate in the contemporary marketplace as "nature under plastic." Most important, Earth Day, which began as a popular green resistance to unfettered capitalist markets, in its mainstream manifestations today has ironically shifted its basic meanings. Increasingly, it takes the form of promoting a very much fettered green capitalism as if it were, despite its artfully managed and slickly packaged commodification, an effective style of meaningful political resistance.

As James Speth, the president of World Resources Institute, observed in 1989 about the first Earth Day in 1970, there has been a "steady and sometimes spectacular growth of worldwide public concern about environmental degradation, and of citizen action and participation to meet these challenges."[46] Perhaps, but such mass-mediated measurements of spectacularly growing concern do not translate into ecological revolution. Earth Days might mark new levels of "intense public interest," as marked by thousands of marchers in Earth Day parades or by pro-environmental sound bites on the nightly news, but, in practice, most consumers' behavior, beyond recycling soda cans or refusing to buy African elephant ivory products, is not radically changing. In part, there only can be minor changes, because change can happen only if the products offered in the marketplace are manufactured in a more environmentally correct manner; and, in part, there will be no radical change, because the broadly mobilized version of green consumerism still is a very passive form of corporate capitalist consumerism.

After nearly two decades of post-Earth Day ecological consciousness, for example, the average per capita daily discard rate of garbage had risen from 2.5 pounds in 1960 to 3.2 pounds in 1970 to 3.6 pounds in 1986.[47] By 2000, despite decades of recycling experience, this figure is expected to rise to 6 pounds a day.[48] Similarly, even though ecological concern is rising, the average gas mileage of new cars declined 4 percent from 1988 to 1990, and the number of miles driven annually has continued to rise by 2 percent every year.[49] Forty-seven thousand square miles of tropical rain forest were cut down in 1979; eighty-eight thousand square miles were cut down in 1989.[50] Japan more than doubled its per capita output of carbon from fossil fuel emissions from 1960 to 1987, Saudi Arabia almost quadrupled its levels, and the United States increased its output by almost 25 percent.[51] Consequently, it becomes apparent that "worldwide public concern" merely may be the contemporary consumer society's Green Cross packaging, wrapped around many of the same old antienvironmental goods and services.

The Ruse of Recycling

As dogmas for attaining not only personal but also planetary salvation, some of today's most common discourses about ecological activism in North America ironically express and unconsciously elaborate a series of unintended consequences, whose unanticipated implications reaffirm tenets of consumption rather than conservation. In the ruse of recycling, green consumerism, rather than leading to the elimination of massive consumption and material waste, appears instead only to be revalorizing the basic premises of material consumption and massive waste. By providing the symbolic and substantive means to rationalize resource use and cloak consumerism in the appearance of ecological activism, the cult of recycling as well as the call of saving the earth are not liberating nature from technological exploitation. On the contrary, they simply are providing a spate of rolling reprieves that cushion, but do not end, the destructive blows of an economy and culture that thrive on transforming the organic order of nature into the inorganic anarchy of capital.

The essential irony of this entire approach to ecology change by green consumerism is that it actually has been at work daily for many years in many millions of households and thousands of firms, at least since the energy crisis of 1973, as a form of do-it-yourself worldwatching that has sustainably developed a green consumer industry. And, after decades of careful ecological concern, more campaigns for recycling, many days of rational shopping, and much thinking about source reduction only have left the biosphere still ravaged by intense ecological exploitation.[52] The earth is not greener or safer, but deader and more endangered. On one level, one must acknowledge that green consumerism actually may have had a slight positive impact on the global environment. After all, and if only for a short time, the planet probably is better off with a few more people using fewer resources at slower rates of consumption. Yet, on another level, these marginal benefits are counterbalanced by the substantial costs of remaining structurally invested in thoroughly consumeristic forms of economy and culture. The "greening" of product advertising, merchandise packaging, or even certain limited technical aspects of the production cycle does nothing to alter the fundamentally antiecological qualities of production in contemporary capitalist society.

This variety of environmentalism is virtually meaningless as a program for radical social transformation because it serves an agenda of conservative ideological containment that also is almost completely anthropocentric. The well-being and survival of other animal species, plant life-forms, or bioregions are virtually ignored. Shoppers for a better world enjoin themselves and others not to buy consumer products made from endangered species or rain-forest beef, but this injunction is driven by other green consumerist needs that parallel these goals. One cannot be a happy ecotourist if there are no longer any rhinos, hippos, or elephants in African game parks, and the rain forests probably contain exotic plants that someday will cure cancers for ailing green consumers. When every group from the Worldwatch Institute to the American Forest Council, Greenpeace to Exxon, the Sierra Club to the Chemical Manufacturers Association claims to be on the same ecological path to green salvation, as today's green consumerism handbooks or Earth Day celebrations indicate, then the most threatening specter haunting the world today is no longer ecologism. Instead, this era of reconciliation shows how decades of rhetorical exorcism directed against radical ecologism have successfully worked their spell by caging this surly antagonistic spirit and refracting its radical revolutionary animus out into the ghostly moonbeams of friendly green consumerism.

7

Marcuse and the Politics of Radical Ecology

Why return to Marcuse? What can his writing possibly offer to those seeking alternatives to the prevailing social order in the 1990s? Since the collapse of the New Left in the 1970s, Marcuse has been left behind as the theory community stampeded from craze to craze during its successive infatuations with Habermas, Foucault, Lyotard, Baudrillard, Derrida, and Heidegger. Marcuse undoubtedly had something to do with this turn of favor after brooding over the demise of New Left radicalism in *Counterrevolution and Revolt*, and then turning away from openly activist strategies to embrace the aesthetic alternatives promised by "a new sensibility" in *The Aesthetic Dimension*.[1] For new audiences caught up in the postmodernism debates of the 1980s, Marcuse's final theoretical stance seemed to lack cultural resonance or political closure.

As a result, his project increasingly was ignored, if not forgotten, by the time of his death in 1979. Up against French fashions in deconstruction, Habermas largely has held his own by taking a much less activist political stance as well as by engaging in debates periodically with the poststructuralists. Yet, in a world that has heard everything Habermas has had to say about philosophical discourses of modernity and theories of communicative interaction, it seems strange that even advanced European societies remain totally bogged down in new crises associated with the end of Nature and tribal wars of fascistic ethnic cleansing. French poststructuralists also may not be of much assistance in this department, but something else is needed beyond Habermas's communicative interaction theories. After trying green consumerism, sustainable development, or worldwatching, something else is needed beyond biospheric engineering or nature conservancies. With regard to the ecological crises posed by the end of Nature, Marcuse still can be quite helpful.

Marcuse and Ecological Issues

As this chapter suggests, Marcuse has always been difficult. His conceptualizations of social contradictions, historical forces, and political conflicts in arcane categories drawn from Freudian metapsychology often lack real complexity. Similarly, his Marxian visions of class domination and Hegelian notions of human needs run against the antifoundationist grain of more recent postmodernist readings, which are highly suspicious of these mystifying metanarratives. Still, Marcuse's critiques are sharp, thorough, and relentless. It is this dimension, particularly with regard to the environmental crises of advanced capitalist society, that remains as vital today as it was three decades ago.

Marcuse's influence among the New Left during the 1960s and 1970s was significant. And, to the extent that elements of the New Left were concerned with issues of ecology and the environment, Marcuse has had a discernible impact on today's ecological criticism. Hazel Henderson's *The Politics of the Solar Age: Alternatives to Economics* marks this facet of Marcuse's work, and Langdon Winner, in *The Whale and the Reactor*, notes how Marcuse, as an ecological thinker, "had begun building a bridge between Frankfurt School critical theory and the possibility of an alternative technology"[2] in the 1960s and 1970s. This side of Marcuse also is noted by Koula Mellos in *Perspectives on Ecology*, who casts him as an important "theoretical inspiration" for the ecology movement through the New Left.[3] Even so, Marcuse rarely is cited as a decisive intellectual influence by contemporary radical ecologists. His project is read now by most as being either so anthropocentric or too socialistic to be taken seriously in the environmental politics of the 1990s.[4]

Ecology and the environment, as they are understood, for example, by today's deep ecologists or bioregionalists, are not prominent features in Marcuse's theoretical project. Like Marx, Marcuse continually makes offhand asides about nature serving essentially as "*man's* inorganic body" in his writings. He affirms the importance of respecting the environment's integrity and order, but he also is committed to rationalizing and humanizing nature. He takes up the topic of ecological destruction per

se only during and after 1970. Even then, his published considerations
are few and unsustained. *Counterrevolution and Revolt*, which was presented initially as lectures at Princeton and the New School for Social
Research during 1970, includes some thoughts on ecology, entitled "Nature and Revolution."[5] Yet, this text was not published until 1972. During the same year, he made some short remarks at a Paris conference on
ecology that also were published in *Liberation* as "Ecology and Revolution" a few months later.[6] Finally, a lecture that Marcuse presented in
California to a group of students during 1979 was published more recently by *Capitalism Nature Socialism* as "Ecology and the Critique of
Modern Society."[7]

  This curious ecological aporia in Marcuse's work can even be docu-
mented indirectly by returning to the two major, book-length analyses
of him published in the United States during the 1980s by Morton
Schoolman and Douglas Kellner. Neither *The Imaginary Witness* (1980)
nor *Herbert Marcuse and the Crisis of Marxism* (1984) specifically identifies the ecology question with Marcuse in its table of contents.[8] Both
studies make no concerted effort to think about or even document Marcuse's approach toward ecology, the environment, or Nature with individual index entries. Marcuse's environmental concerns occasionally are
raised for discussion by both Schoolman and Kellner, but neither develops a truly concentrated elaboration of Marcuse's views on environmental topics.

  On the one hand, this silence is understandable. In spite of his repu-
tation for being the all-knowing guru of the New Left, Marcuse does
not consider the ecological problematic as such until other figures and
forces associated with New Left movements popularized these questions
during the months leading up to the first Earth Day in 1970. Even then,
the environmental question remains tied up within Marcuse's Marxian
reading of Nature and Freudian approach to human subjectivity. On the
other hand, however, much of Marcuse's theoretical project does focus
on ecology and the environment. His most important work assesses the
negative impact of excessively destructive social institutions on "human
nature," or the primary impulses and experiences underlying anyone's
rationality and emotions, and "external nature," or the existential environments of Nature that frame everyone's survival. These preoccupations

are central to his analysis of domination in *Eros and Civilization, One Dimensional Man,* and *An Essay on Liberation.*[9] In *Counterrevolution and Revolt,* for example, Marcuse asserts that "in the established society, nature itself, ever more effectively controlled, has in turn become another dimension for the control of man: the extended arm of society and its power."[10] Consequently, the revolutionary problematic of the present era is quite clear: "the radical transformation of nature becomes an integral part of the radical transformation of society."[11]

The Ecological Dimension in Marcuse's Project

Like many of today's radical ecologists, Marcuse argues in *One Dimensional Man* that "contemporary industrial society tends to be totalitarian."[12] Totalitarian forms of rule include not only terroristic, one-party dictatorships, but also any ecological social formation or psychosocial production regime that can be tied to "a non-terroristic economic-technical coordination which operates through the manipulation of needs by vested interests."[13] Marcuse admits that the character, satisfaction, and intensity of human needs always have been historically preconditioned, and that the question of what are true and false needs ultimately can be answered only by the individuals expressing these needs. Under late capitalism, the sociohistorical definition of needs, the politico-economic structures that promote the repressive development of individual needs, and the technical-administrative system for satisfying socially defined/personally accepted needs all must be "subject to overriding critical standards."[14]

Subjectivity and Productivity

Marcuse's critique of advanced industrial society explores the most perplexing issue posed by the ruses of technological reason, namely, "how can civilization freely generate freedom, when unfreedom has become part and parcel of the mental apparatus?"[15] Everything in society must be gauged by how much actual freedom from material want is becoming a real possibility. Under these conditions, Marcuse never ceased believing in the utopian hopes that Marx placed on making the leap from the realm of necessity into the realm of freedom:

> The very structure of human existence would be altered; the individual would be liberated from the work world's imposing on him alien

needs and alien possibilities. The individual would be free to exert autonomy over a life that would be his own. If the productive apparatus could be organized and directed toward the satisfaction of the vital needs, its control might well be centralized; such control would not prevent individual autonomy, but render it possible. This is the goal within the capabilities of advanced industrial civilization, the "end" of technological rationality.[16]

Although these emancipatory promises actually are possible for Marcuse, they are not being realized. Vested interests, which survive by controlling the state, the productive apparatus, and the institutions of society, manipulate psychosocial expectations with strategies of repressive normalization that impose false needs on individuals and collectivities. "Such needs," Marcuse notes, "have a societal content and function which are determined by external powers over which the individual has no control; the development and satisfaction of these needs is heteronomous."[17] "True needs," as opposed to "false needs," are those vital human needs for food, lodging, clothing, and meaning at some ecologically sustainable level of culture.

In keeping with the critique advanced by many radical ecologists, Marcuse attacks "false needs," or "those which are superimposed upon the individual by particular social interests in his repression: the needs which perpetuate toil, aggressiveness, misery, and injustice."[18] As Marcuse notes,

> their satisfaction might be most gratifying to the individual, but this happiness is not a condition which has to be maintained and protected if it serves to arrest the development of the ability (his own and others) to recognize the disease of the whole and grasp the chances of curing the disease. The result then is euphoria in unhappiness. Most of the prevailing needs to relax, to have fun, to behave and consume in accordance with the advertisements, to love and hate what others love and hate, belong to this category of false needs.[19]

With these arguments about individual subjectivity and social productivity, Marcuse presents a comprehensive analysis of how and why an advanced industrial society functions on antiecological terms. By exploiting Nature, it produces a short-range, material surplus that allows its vested controlling interests to co-opt, buy off, or immobilize "those needs which demand liberation—liberation also from that which is tolerable and rewarding and comfortable—while it sustains and absolves the destructive power and repressive function of the affluent society."[20] Everyday material existence can be quite tolerable, rewarding, and comfortable because it requires deep, long-run, ecological disaster to sustain its shallow, short-run institutional reproduction. False needs become the cause of and excuse for continuing such environmental destruction as everyday life merely vindicates "the freedom to choose." What is chosen, however, is the perpetuation of a false repressive totality where liberty is transformed into happily accepting the mechanisms of domination. "Here," as Marcuse observes,

> the social controls exact the overwhelming need for the production and consumption of waste; the need for stupefying work where it is no longer a real necessity; the need for modes of relaxation which soothe and prolong the stupefication; the need for maintaining such deceptive liberties as free competition at administered prices, a free press which censors itself, free choice between brands and gadgets.[21]

This waste represents not only serious social irrationality, but also complete environmental disaster.

Marcuse's understanding of the ecological crisis is closely tied to his reading of subjectivity through the optics of Freudian metapsychology in which human beings are shaped by two basic instincts. One is Eros, or erotic energy and the life instincts; the other is Thanatos, or destructive energy and the death instincts. Unfortunately, the major reality principles of advanced industrial society, or the sum total of those norms and values that regulate moral behavior, are based on the destructive energies of Thanatos. The death instincts of Thanatos express a human drive to live in a state of painlessness in the womb prior to birth. Its force, as Marcuse argues, "is the destruction of other living things, of other living beings, and of nature."[22] These drives are at the heart of the repressive false needs in one-dimensional society, and they anchor the performance principles of toil and sacrifice at the core of technological rationality. To oppose its workings, Marcuse looks to Eros, or to the life instincts, for the basis of resisting this destructive social order. This drive, according to Marcuse, pushes to attain not the painlessness prior to the beginning of life but the full, flowering maturity of life: "It would serve to protect and enhance life itself. The drive for painlessness, for the pacification of existence, would then seek fulfillment in protective care for living things. It would find fulfillment in the recapture and restoration of our life environment, and in the restoration of nature, both external and within human beings."[23] The constellation of false needs presented to the inhabitants of advanced industrial society creates a conformist character structure, while, at the same time, blocking the emergence of a radical character structure that might transform it by reopening human subjectivity to Nature. The radical character structure threatens this entire social order, because in looking to natural forces it represents "a preponderance in the individual of life instincts over the death instinct, a preponderance of erotic energy over destructive drives."[24] Given this organic basis for radical subjectivity, Marcuse connects the liberatory agendas of the ecology movement to the expression of Eros as an organized political force: "This is the way in which I view today's environmental movement, today's ecology movement. . . . A successful environmentalism will, within individuals, subordinate destructive energy to erotic energy."[25] Various contemporary ecology movements also might be understood as moves to politicize such erotic energy, even though they now lack the institutional power to overthrow the ruling reality principle. Even though their rank and file may express a desire for radical change, these diversely divided movements remain bogged down by ineffectual strategies for organizing nonconformist, protest campaigns rather than striking out to totally reconstitute society from the ground up.

Technology and Ecology

Marcuse's reading of science and technology in one-dimensional society rearticulates much of the Frankfurt School's critique of the Enlightenment.[26] Ultimately, Marcuse sees science, as it operates in contemporary advanced industrial society, in terms that underscore its intrinsic instrumentalism. The procedures of abstraction, calculation, formalization, and operationalization lead him to contest "the *internal* instrumentalist character of this scientific rationality by virtue of which it is *a priori* technology, and the *a priori* of a specific technology—namely, technology as a form of social control and domination."[27] This inherent instrumentalism is problematic, because the value-free objectivism of science leaves it open to adopt and serve substantive ends that are external to it. Emerging along with modern European entrepreneurial capitalism and nationalistic statism in Europe, the inherent technological instrumentalism of science soon integrated destructive social ends into its operations.

> The principles of modern science were *a priori* structured in such a way that they could serve as conceptual instruments for a universe of self-propelling, productive control; theoretical operationalism came to correspond to practical operationalism. The scientific method which led to the ever-more-effective domination of nature thus came to provide the pure concepts as well as the instrumentalities for the ever-more-effective domination of man by man *through* the domination of nature. Theoretical reason, remaining pure and neutral, entered into the service of practical reason. The merger proved beneficial to both. Today, domination perpetuates and extends itself not only through technology but *as* technology, and the latter provides the great legitimation of the expanding political power, which absorbs all spheres of culture.[28]

Caught up within these operational constraints and goals, science works so that "the liberating force of technology—the instrumentalization of things—turns into a fetter of liberation; the instrumentalization of man."[29]

Humanity's increasing control over the environments of Nature through technological means necessarily results in a greatly increased ability to dominate human nature. The two spheres are intimately connected inasmuch as the complex technical controls implicit in advanced technology demand that everyone exercise greater discipline over his or her own labor and patterns of consumption. By preconditioning the behavior of individuals, Marcuse sees technological reason introjecting its technical demands into each person's somatic-psychic constitution, which "becomes the psychological basis of a *threefold domination*: First, domination over one's self, over one's nature, over the sensual drives that want only pleasure and gratification; second, domination of the labor achieved by such disciplined and controlled individuals; and third, domination of outward nature, science, and technology."[30]

Science and technology become an antienvironmental system of domination with its own subpolitics of instrumental control. This recognition is critical:

> Science, *by virtue of its own method* and concepts, has projected and promoted a universe in which the domination of nature has remained linked to domination of man—a link which tends to be fatal to this universe as a whole. Nature, scientifically comprehended and mastered, reappears in the technical apparatus of production and destruction which sustains and improves the life of individuals while subordinating them to the masters of the apparatus.[31]

Consequently, a rationalizing technical hierarchy based on humans dominating Nature merges with a disciplinary social hierarchy of humans dominating other humans in the abstract machinery of one-dimensional society.

Marcuse also sees a possibility for changing the negative trends in the scientific project. The reconciliation of science and technology as a global system, or Logos, within a new metaphysics of liberation, or Eros, might assist science in developing essentially different concepts of nature, facts, and experimental context. Beyond the reification of technology, which reduces humans and Nature to fungible objects of organization, neither the worlds of Nature nor the systems of society would be the stuff of total administration. Marcuse believes that this break would be possible if a new idea of Reason attuned to a new sensibility capable of guiding its theoretical and practical workings could be developed. This moment, which would reverse the relationship between existing science and a metaphysics of domination, would come with the completion of technological rationalization, or "the mechanization of all socially necessary but individually repressive labor."[32] This moment of technological liberation also would make possible the pacification of existence—a new social condition marked by qualitatively different relations between humans as well as between humans and nature—if such newly freed individuals would work effectively to finally realize this emancipatory moment.

The "New Sensibility" and Pacifying Nature

Marcuse's ecological engagement is intertwined with his search for "a new science" and "a new sensibility" as paths for society to take itself out of its current environmental crises. If the old science of instrumental operationalism is behind the domination of Nature and humanity in the abstract machines of industrial environments, then new scientific practices, linked not to a metaphysics of domination but rather to a metaphysics of liberation, might well alter everything. Here, a new sensibility—aesthetic, life-affirming, and liberatory in character—would play a vital role. Most important, a new sensibility, based on aesthetic dimensions and a regard for beauty as a check against aggression and destruction, would mark the ascent of Eros over Thanatos in the pacification of existence.

Marcuse sees the powers of the imagination, unifying the faculties of sensibility and reason, becoming productive and practical. A new sensibility of emancipatory freedom would work as "a guiding force in the reconstruction of reality—reconstruction with the help of a *gaya scienza*, a science and technology released from their service to destruction and exploitation, and thus free for the liberating exigencies of the imagination."[33] The new science, when combined with the sensuous aesthetic awareness of the new sensibility, could reintegrate labor and leisure, science and art, work and play so thoroughly that humanity and nature would also become one: "such a world could (in a literal sense) embody, incorporate, the human faculties and desires to such an extent that they appear as part of the objective determinism of nature."[34] By unchaining reason from domination, and exalting Eros over Thanatos, humans with the new sensibility would mobilize the aesthetic to develop freedom hand in hand with emancipation as art merges with technology, and science serves liberation.

> The aesthetic universe is the *Lebenswelt* on which the needs and faculties of freedom depend for their liberation. They cannot develop in an environment shaped by and for aggressive impulses, nor can they be envisaged as the mere effect of a new set of social institutions. They can emerge only in the collective *practice of creating an environment*: level by level, step by step—in the material and intellectual production, an environment in which the nonaggressive, erotic, receptive faculties of man, in harmony with the consciousness of freedom, strive for the pacification of man and nature. In the reconstruction of society for the attainment of this goal, reality altogether would assume

a *Form* expressive of the new goal. The essentially aesthetic quality of this Form would make it a work of *art,* but inasmuch as the Form is to emerge in the social process of production, art would have changed its traditional locus and function in society: it would have become a productive force in the material as well as cultural transformation.[35]

Art would cancel the deadening operational facticity of technological domination in today's abstract machines with its enlivening visions of technological emancipation in tomorrow's tangible communities. In the development of society and the subject, Marcuse argues, the human pacification of existence can be repressive or liberating. Nature is not seen as some benevolent, all-knowing fount of positive goodness; it is instead constructed by Marcuse as a combination of ferocious, inventive, blind, fertile, and destructive processes. And, the liberating pacification of Nature should reduce the misery, violence, and cruelty of Nature in the face of its scarcity, suffering, and want.

"Nature and Revolution," in *Counterrevolution and Revolt,* brings Marcuse directly to ecology and the environment through his "new sensibility."[36] Now trapped by psychosocial performance principles no longer needed to produce the material foundations of civilization, Marcuse sees individuals as having fresh chances for attaining liberation by developing intentionally new sensibilities about the unlimited liberatory potential of all modern technology. On this count, Marcuse asks Freud for some preliminary directions, but he does not accept Freud unquestioningly as an all-knowing guide in these murky realms of analysis. For advanced industrial society, Marcuse argues, "the performance principle enforces an integrated repressive organization of sexuality and of the destruction instinct."[37] However, if the unintended consequences of technological rationalization have rendered the institutions of the performance principle obsolete, then "it would also tend to make obsolete the organization of the instincts—that is to say, to release the instincts from the constraints and aversions required by the performance principle."[38]

This claim could imply, on the one hand, the eventual elimination of such destructive surplus repression in new emancipatory forms of life; or, on the other hand, it might explain why ruling social forces generate false needs by adhering to the performance principle long after it has ceased being needed to meet vital needs. To transcend the performance principle of advanced capitalism, Marcuse believes, "individuals themselves must change in their very instincts and sensibilities if they are to build, in association, a *qualitatively* different society."[39] These changes require not only the emancipation of consciousness, but also the emancipation of the senses, to envelope the totality of human existence.

At the heart of his new sensibility, Marcuse affirms Marx's vision for transforming society. Only now he sees a revolution to be made in accordance with "laws of beauty" by underscoring the importance of aesthetic needs and impulses. In reversing capitalism's repressive containment of the aesthetic dimension in performative laws of profit, and redirecting aesthetic awareness as a subversive force, Marcuse wants the active/aggressive destructiveness of capitalism to be upended by the passive/receptive productiveness of a new socialist community. This outcome would, in part, reflect the unleashing of more positive, but repressed and distorted, "female" qualities to recombine with the negative, but also oppressive and contorted, "male" qualities. Ultimately, what Marcuse wants to see come into realization here is "the ascent of *Eros* over aggression, in men *and* women; and this means, in a male-dominated civilization, the 'femalization' of the male. It would express the decisive change in the instinctual structure; the weakening of primary aggressiveness which, by a combination of biological and social factors, has governed patriarchical culture."[40]

These shifts might begin to emancipate Nature from the exploitative domination of destructive technologies. And with it would come "the ability to see things in their own right, to experience the joy enclosed in them, the erotic energy of nature—an energy which is there to be liberation; Nature, too, awaits the revolution."[41] Human emancipation would also entail Nature's becoming expressive of human historical qualities. With the fusion of Eros with *technē*, Marcuse argues that new aesthetic realizations should take place, although he is unclear about what the aesthetic register ultimately is. This revolutionization by aesthetic means also would bring with it a new ecological order. On the one hand, "cultivation of the soil is qualitatively different from destruction of the soil, extraction of natural resources from wasteful deforestation;

and, on the other hand, poverty, disease, and cancerous growth are natural as well as human ills—their reduction and removal is liberation of life."[42] The pacification of existence would be a truly postmodern condition in which modern, aggressive, technological society no longer struggles to dominate and exploit Nature. Instead, it becomes fully humanized, civilized, pacified in the conquest of necessity; thus, "Nature ceases to be mere Nature to the degree to which the structure of blind forces is comprehended, and mastered in the light of freedom."[43]

Marcuse's ecological awareness helps him to see how the technological means to conquer scarcity also have turned into tools for forestalling liberation. Obscene levels of overproduction and excessive consumption enjoyed in many advanced industrial regions cannot provide an acceptable model for the pacification of existence, because they are accompanied "by moronization, the perpetuation of toil, and the promotion of frustration."[44] The environment is plundered to provide the materials needed for the one-dimensional society; and, "it is the sheer *quantity* of goods, services, work, and recreation in the overdeveloped countries which effectuates this containment. Consequently, qualitative change seems to presuppose a *quantitative* change in the advanced standard of living, namely, *reduction of overdevelopment*."[45] Only the existing technical base of overdevelopment in advanced industrial society can provide the rational foundations for beginning the pacification of existence; but it is this same power that perpetuates the dehumanizing tendencies of one-dimensional society.

Marcuse's program for pacifying Nature is neither ridiculous nor impossible. His vision is fragmentary and incomplete, but he outlines an agenda in concrete political terms. In contrast to one-dimensional society, marked by "the increasing irrationality of the whole; waste and restriction of productivity; the need for aggressive expansion; the constant threat of war; intensified exploitation; dehumanization," Marcuse chooses to pursue an alternative, rooted in "the planned utilization of resources for the satisfaction of vital needs with a minimum of toil, the transformation of leisure into free time, the pacification of the struggle for existence."[46] Unlike many of today's ecofeminists or deep ecologists, who zoom around the planet on jumbo jets burning tons of fuel to decry the pollution of the atmosphere, the evils of modern technology, and corruptions of consumerism, Marcuse is more honest about his vision of pacifying Nature. Since nature is a human construct in both theory and practice, truly nonanthropocentric society or posttechnological economy is pure fantasy. Since the pacification of Nature presupposes a partial mastery of Nature, which is and remains the impassive objectivity opposed to the formation of liberating institutions, a new science would need its guiding illusions from a new sensibility grounded on art. At this juncture, "the rationality of art, its ability to 'project' existence, to define yet unrealized possibilities could then be envisaged as *validated by and functioning in the scientific-technological transformation of the world.* Rather than being the handmaiden of the established apparatus, beautifying its business and its misery, art would become a technique for destroying this business and this misery."[47]

Marcuse and Ecological Criticism Now

Today's ecology and environmental movements are complex, diverse, and pluralistic. Ideas that influence one faction, such as animal rights philosophy, ecological economics, deep ecology thinking, or global energy accounting, often are completely disdained or wholly ignored by other groups in what most outsiders would regard as different branches of the same basic movement. Marcuse's influence on any one faction of these environmental movements is difficult to document, even though his ideas parallel many positions adopted by many elements in the ecology and environmental movements. In the 1960s, neither Barry Commoner nor Murray Bookchin, for example, gives much indication of being influenced directly by Marcuse in his work, although Bookchin's *The Ecology of Freedom* mocks Marcuse's visions for realizing the pacification of Nature.[48] At the same time, somewhat more conventional readings of ecological crises developed by Rachel Carson, Herman Daly, and David Brower do not acknowledge Marcuse.[49]

In the 1970s and 1980s, new ecological thinkers, such as Arne Naess, Bill Devall, George Sessions, E. F. Schumacher, Dave Foreman, Ivan Illich, Thomas Berry, Carolyn Merchant, Henryk Skolimowski, Wendell Berry, Bill McKibben, and Kirkpatrick Sale, similarly give little sense of being influenced by Marcuse.[50] Of the three major histories of the ecology and environmental movement, either published or revised in the

1980s—Roderick Nash's *Wilderness and the American Mind,* Samuel Hays's *Beauty, Health, and Permanence: Environmental Politics in the United States, 1955–1985,* and Anna Bramwell's *Ecology in the 20th Century,* only Bramwell even mentions Marcuse, and then it is mainly in passing when discussing the New Left of the 1960s.[51] Still, Donald Edward Davis includes Marcuse's *One Dimensional Man* in his 1989 overview of ecological thought, *Ecophilosophy: A Field Guide to the Literature,* as an important influence on ecological philosophers and environmentalist thinkers.[52]

Marcuse's theories about technology, subjectivity, and Nature are not without problems. His quest to discover new organic sources of social negativity and political resistance in late capitalism ultimately led him through classical Marxism to Heidegger and Freud. This search culminates, in turn, with his phenomenological critique of technological rationality and psychoanalytic theory of history. Marcuse's adaptation of all these varying perspectives resulted in some problematic misrepresentations of the emancipatory possibilities in present-day political realities. Marcuse also perhaps proved insufficiently critical of technological rationality as he attributed its domination largely to its misuse by exploitative groups. Similarly, he ends his critique of modern technological society by grounding his emancipatory politics for a new ecological collective subjectivity in the organic instinctual energies of each human individual. Searching to supplant the historical negativity of the identical subject-object of *labor,* or the emancipated proletariat, Marcuse adopts an equally problematic solution, namely, a new naturalistic, presocial, and prehistorical collective subjectivity—the identical subject-object of *pleasure,* or the individual's and the human species' erotic instincts.[53]

Still, Marcuse cannot be easily dismissed. He anticipates virtually every critique made by contemporary radical ecology groups. First, like deep ecology, he identifies the destruction of nature with instrumental reason, or "a concept of reason which contains the domineering features of the performance principle"[54] in order to ground all of his ecological arguments. Second, like ecofeminism, he connects the workings of the performance principle with the destructive drives of the death instinct and male needs for domination. Third, like social ecology, he sees the domination of Nature flowing out of the domination of human beings inasmuch as ruling forces and vested interests in society subject internal human nature and external environmental nature to the same destructive forms of domination. Fourth, like many soft path technologists, he suggests that modern technology possesses the power and productivity to overcome material scarcity, if only its techniques and instrumentalities were organized in more rational emancipatory forms of application. Fifth, like advocates of voluntary simplicity, he ties waste, ruin, and despoliation of the environment to false needs imposed on individuals not to meet true vital requirements but to perpetuate the powers and privileges of vested interests that benefit from such domination and destruction. Finally, like the new nature poets and philosophers, Marcuse expects a new sensibility—one that is life-affirming, aesthetic, female, erotic, and liberatory—to provide the conceptual categories and moral values needed to reintegrate humanity with Nature in an environmentally rational society where technology is art, work can be play, and ecology provides freedom.[55]

Despite his shortcomings, then, Marcuse is a theoretical force to be reckoned with. Much of today's debate within deep ecology, ecofeminism, social ecology, and bioregionalism merely revisits issues of political conflict, cultural contradiction, and individual struggle that Marcuse first raised in *Eros and Civilization, One Dimensional Man, An Essay on Liberation,* and *Counterrevolution and Revolt.* The question of a new science, a new technology, and a new aesthetics as the basis for realizing an ecological transformation of society has yet to be addressed as sharply as Marcuse did, even if his critical vision of these forces has problematic qualities. Soleri and Bookchin, in their own ways, as chapters 8 and 9 show, begin to examine how a new science/technology/aesthetics could transform the workings of today's ecologically destructive megamachineries. For this reason alone, Marcuse's work needs to be read closely again, because his visions of a pacified existence for humanity in Nature often seem far more plausible than the dour green visions of planetary reengineering presented by today's ecoauthoritarians.[56]

8

Developing an Arcological Politics:
Paolo Soleri on Ecology, Architecture,
and Society

Atop a low mesa above the Agua Fria river in central Arizona, a unique laboratory devoted to testing a new ecological alternative has been developing slowly for twenty-five years. A project of Paolo Soleri's Cosanti Foundation, Arcosanti is a working prototype for a new kind of city, one that is being designed, built, and inhabited as a three-dimensional, highly concentrated megastructure.[1] To house a city of five thousand to six thousand people, Arcosanti will occupy only fifteen acres of land in the midst of an 860-acre greenbelt/park area/agricultural zone. The closely articulated structures of Arcosanti will be not much more than one-quarter of a mile on any one side, but they will rise to as much as thirty stories tall. Within them, Soleri and his coworkers will locate the economic and social infrastructure of any modern city, while providing residents up to two thousand square feet of living space per family to do with more or less as they please. Outside, everyone will be able to enjoy the expansive views of another three thousand acres to be kept as undeveloped open space.

Soleri has used Arcosanti to rethink the totality of modern urban design and planning as his response to the environmental crisis. Instead of accepting the logic of explosion undergirding today's two-dimensional cities, such as Phoenix or Los Angeles, which follow an automobile-driven "scatterization" pattern by pushing further and further outward from their peripheries as their inner cores atrophy and die, Arcosanti is a laboratory for urban implosion.[2] Soleri's unusual fusion of architecture with ecology in what he calls "arcology" bans the automobile from operating within the city in favor of pedestrian walkways aided by fixed me-

153

chanical people movers (elevators, escalators, and moving sidewalks).[3] In a city of such complex compactness, most journeys by foot would require no more than fifteen or twenty minutes—about the same amount of time it takes to walk from inside a major mall to the outer ring of the parking lot in a modern metroplex such as Phoenix or Los Angeles. In staging this controlled implosion of an American city, Soleri will stabilize his community at about 350 people per acre—ten times the population density of New York City. Its ecological superiority derives from the source of its implosion—the elimination of the automobile, and hence all of the associated space costs charged off to cars and trucks in streets, highways, parking, dealerships, fueling, repairs, and junkyards in order to provide the "benefits" of an automotive transportation system.[4]

By rethinking the twentieth-century industrial city from its underground infrastructure up to the ground level and then on out to the sky, Soleri's fusion of high technology with ecological responsibility in the arcology gives us one of the more realistic responses—despite Arcosanti's being frequently derided as the misbegotten folly of a utopian dreamer—available right now as a communitarian answer to the environmental crisis. This effort is not merely a paper proposal; Soleri's arcological alternative is being built and lived in today. Despite all its promise, however, much of Arcosanti's potential is going unfulfilled. There are many unresolved problems in Soleri's ecological project, casting doubt over how successfully it has been implemented in Arizona since the early 1970s. This chapter addresses some of these concerns, but does so recognizing that most of them arise because Arcosanti is an already working prototype, not simply some inoperable dream.

Soleri's Career

Paolo Soleri was born in Turin, Italy, in 1919. Educated at the Turin Polytechnical Institute, where he took a doctorate in architecture, he joined Frank Lloyd Wright at Taliesin West outside of Phoenix, Arizona, in 1948–49. After his Wright apprenticeship in 1949, he, his wife Corolyn, and Mark Mills built the "Dome House" near Cave Creek, Arizona—a structure that was closely integrated into its natural environment in accord with Wright's organic architecture theories, but that also is filled with intimations of Soleri's later work. In 1950, Soleri returned to Italy and settled in Vietri, where he built a house and studio. Soon he was commissioned by the Solimene family to design a new ceramics factory, which interested him in ceramics manufacturing processes. Once this plant was completed, Soleri and his wife moved back to Arizona in 1955, and bought five isolated acres of land in Paradise Valley outside of Phoenix (about fifteen miles southwest of Wright's Taliesin West), where they began to produce ceramic and bronze objects, including the now world-renowned wind-bells and wind chimes that, over the intervening decades, have provided most of the funds for Arcosanti's modest yearly construction budget. In 1956, he also established the Cosanti Foundation as a studio, shop, and residence, first, to produce his ceramics, and, then later, to illustrate on a very small scale the functional utility of his architectural ideas.[5]

Throughout the 1950s and 1960s, Soleri worked through his maverick approaches to reengineering the material interfaces between the environment and society while watching the Phoenix metroplex rapidly encroach upon his desert homesite. After gaining some level of notoriety in the 1960s as an ethical philosopher, social critic, and architectural visionary, for which he won a Guggenheim Fellowship in 1965 to plan an entire arcological urbanism called Mesa City, he began building the Arcosanti megastructure at Cordes Junction, Arizona, in 1970. This project, along with his writing and educational work, has occupied him almost totally over the past quarter century.

The ceramics connection is not an insignificant aspect of Soleri's architectural thinking. Casting ceramic and bronze items provided Soleri with much of his income during these early years, but the techniques used in casting ceramic artifacts also moved him to experiment with them as architectural construction techniques:

> I began working with earth and silt in the early 1950s. Originally I became interested in using desert soil and the silt abundant in the dry Arizona river beds because of their inherent properties and availability. Experimentation proved the usefulness of earth and silt as molding mediums for many types of crafts projects. Clay and plaster were the first materials that we cast in earth or silt molds. We produced ceramic windbells from earth molds, and plaster architectural models

from originals which had been carved in silt. The use of earth and silt
for making forms on which to cast concrete was the logical next step.[6]

From 1962 to 1974, six of the structures at Cosanti were "earth cast"
using river silt forms and involving architecture students interested in
learning these unusual building techniques. "Today these structures,"
Soleri observes, "comprise the Cosanti complex, and are used as craft
studios, work areas, offices, and residences. The Cosanti complex is the
result of a combination of ancient craft techniques, new variations on
these techniques, scrounged and donated materials, aesthetic perceptions, unorthodox architectural concepts, and the sweat of many workers."[7] The workshops moved out to the Arcosanti site during the 1970s,
and much of that structure also has been built with casting techniques
extrapolated from the Cosanti ceramic production process.

Most of Soleri's writings were produced in the 1960s and 1970s
when he, unfortunately or fortunately, chose to engage many of the pop
icons of that era—Marshall McLuhan, R. Buckminster Fuller, Teilhard
de Chardin, Woodstock Nation, transcendental meditation, hippie
communes—in his discussions of arcology.[8] While Soleri never attained
the same celebrity status as some of these figures, his writings, especially
when read today, often seem anachronistic. The Teilhardian tone can be
baffling, even though it is central to his philosophies. Nonetheless, the
importance of Soleri's underlying ideas is apparent; much of what he
wrote two or three decades ago is only more true today, even though
the frames of reference for staging its arguments seem fairly antiquated.

Soleri is not a thinker for cynical postmodernists. The rhetorics of
his thinking operate in rigid oppositions and hard contradictions: matter and spirit, process and growth, scatterization and complexification, Alpha God and Omega God, giganticism and miniaturization. For
Soleri, what is and what ought to be are clearly knowable. Architecture,
as a mediation of "theotechnology" to realize "the urban effect" out of
chaotic Nature, is a powerful means for transforming humankind to realize its inner cosmic mission. Such certitudes rankle some postmodern
sensibilities, but Soleri is unabashedly anthropocentric in almost every
one of his writings. Even so, he needs to be considered seriously, because
his environmental analysis could greatly improve radical ecology's intel-

lectual mission of world disclosure. Few more conventional environmental thinkers have contributed as much as Soleri to understanding how the built environment, Nature, social ecologies, or the city actually operate under advanced industrial conditions. Soleri asserts: "the most common mistake about my work is the belief that some years of introspection have produced a take-it-or-leave-it package. . . . rather I am proposing a methodology and at the same time trying to illustrate it."[9]

Soleri's Project

Architecture for Soleri is fundamentally a social calling. Most significantly, it can lead an ecological revolution, because it is an informational process rather than a material activity. Each edifice in the man-made environment ultimately operates as an actor, "that is to say, it is actually information in and of itself instead of being a remote presence."[10] And, this embodied agency pertains not only to the architecture of individual buildings, but also to all architecture seen as the combination of all built environments, all artificial landscapes, or the engineered infrastructures of both urban and rural regions. When viewed from this perspective, architecture can be apprehended as much more than mere construction in its total ecological impact:

> Architecture embraces virtually all of the nondisposable, noninstant, and relatively nonobsolete world. Architecture is the alterations made upon nature by the organic, the psychological, the mental, the components of man's consciousness where the social-cultural stresses operate within and emanate out of the human kind. Architecture . . . is not only a shelter for communication and information institutions, a medium, but it is also . . . mass information itself.[11]

Architecturalized environments, then, materially express and embody what each society does. Since architecture-as-information actively is involved in forming each personality and community, Soleri concludes, "it is then only logical that the pauperization of our soul and the soul of society coincide with the pauperization of the environment. One is the cause and reflection of the other."[12] Any radical transformation in society also demands that one reorder not only ideas and institutions but also society's most basic informational technology: the architectures of its built environment. As Soleri sees ecology, environmentally sound development projects must be rethought as conjoined pieces of humanity's theological and technological evolution. Organizing spaces and designing places, which is the architect's task, is an enterprise invested in "the business of sacredness." Although the urban metroplexes of today fall far short of this task, "the discrepancy is to be seen as a gap between what ought to be and what is," and, correct ecological design can be directed to attain "the *Civitas Dei*. It is that complex machine for information which by the nonexpedient ways of design also becomes knowledge in itself, where the media has finally, if only within limits, sublimated itself into the message, a large interiority reverentially turned into the handling of *particulae sacrae*, the polis dwellers."[13]

Soleri freely mingles elements of theological speculation, moral outrage, construction theory, and eschatological design in his writings to advance what he regards as his "metaschematics for cosmogenesis." Arcologies, such as Arcosanti, represent much more than mere experimentations with alternative urban technology or subversive communitarian resistance. Arcosanti is both of these, but those engagements are more secondary to what Soleri sees primarily as an ethical struggle for the concrete enactment of spiritual enlightenment. His architectural ideas constitute a technology for the immanent realization of an emergent divinity from a volatile universe of emergence: "we can say that the true God is *not yet*, but the true God *will be*, because the creational process is."[14]

Such assertions already violate many modern ontological assumptions: matter is not dead, spirit is not merely transcendent, the universe is not a finished process, man is not a machine, god is not an abstraction. Soleri's cosmological premises are openly derived from Teilhard de Chardin's odd pantheistic musings.[15] The fine points of Soleri's engagement with Teilhard's philosophy, however, do not need to be discussed here. Although he clearly disagrees with Teilhard at several junctions, Soleri does see arcology as expressing his eschatological thinking. More concretely, "Arcosanti tries to 'radicalize' the proposition by attempting to reinstate the working presence of complexity-miniaturization, quite probably in the best 'Teilhardian tradition.' Therefore, the 'how-ness' of Arcosanti is the quest of a more fitting instrument for the human animal to go on in the transcendental quest for grace, also, I think, a Teil-hardian aim."[16]

Ecology, then, becomes humanity's most crucial science inasmuch as the ecosystems of Earth are mind-generating, spirit-attaining processes in which matter will become spirit. Aesthetic practices, or "esthetogenesis," are a key mediation of this dialectic. As Soleri exclaims, art is the center of an ecology-environment-theology triad:

1. The environment: immanent manifestation of an eschatological drive.
2. Religion: simulation model of anticipated perfection, space, and the fully aware divinity of a concluded reality.
3. The aesthetic: emergence of the spirit from matter in specific particles of grace (a novel, a dance, and so on).[17]

Aesthetic imagination and artistic discipline in many ways become the vital spark, wholly dependent on individual labor and communal energy, animating matter with spiritual significance.

Soleri's thinking about "the urban effect" as *homo faber* fused with *cosmos faber* in a city's aesthetic creation grounds itself on the "complexity-miniaturization-duration imperative." More concretely, he sees all of Nature, "from bacteria to God," conforming to three fundamental principles:

1. COMPLEXITY. Many events and processes cluster wherever a living process is going on. The make-up of the process is immensely complex and ever intensifying.
2. MINIATURIZATION. The nature of complexity demands the rigorous utilization of all resources—mass-energy and space-time, for example. Therefore, whenever complexity is at work, miniaturization is mandated and a part of the process.
3. DURATION. Process implies extension of time. Temporal extension is warped by living stuff into acts of duration. A possible resolution of "living time" is the metamorphosis of time into pure duration, i.e., the eventual "living outside of time."[18]

This idiosyncratic reading of the cosmos provides a nomological code that Soleri regards as "a clear, forceful normative light for any living process to follow."[19] There is an immanent order in Nature that human reason and imagination can discover. Through arcological design, Arcosanti's construction is being organized with an awareness of these interlocked norms, because all of "Nature and the living are dependent on such coherence."[20]

Contrasting the simple sprawl of lower life forms (coral, bacteria, mold) to the complex concentration of higher life forms (bees, wasps, ants) in Nature, Soleri regards the most successful and sophisticated forms of life, such as the human city, as those allowing complexity/miniaturization/duration to coalesce. Cities are, in fact, expressions of genuine ecological balance between city and country, even though these relations are becoming attenuated and/or broken in the post-World War II era as noncity/noncountry exurban and suburban spaces proliferate as more simple, more hypertrophic, and more transitory sprawls. Sprawl unfolds on an inhuman scale, because it is fabricated for and by the automobile. This abuse of space is the sociopathic condition that Soleri seeks to cure; such metropolitan sprawl generates "non-structured nomadism" that simultaneously finds "liberation of a sort" in a market-based "license to abuse the environment."[21]

Technology jibes fully with cosmogenesis, and Soleri's own arcological techniques fulfill an eschatological assignment. An arcology, or "ecological city," is one of these vital post-biological technologies:

> It is a complex instrument, but it is a medium which is in itself also a message, inasmuch as it is tangible and pragmatically part of the transpersonal nature of life (that part which is of a theological nature). In fact, ultimately Arcology must be stone made into spirit or it will be a simple mechanism of economic, political and logistical expediency. It has to be the spirit moving the mountains not so much because of the physical manipulation of the mountain (mining, etc.) but because of an inner flame elicited within the stone itself by design and grace . . . the process of esthetogenesis. . . . for this to be so it suffices that first, "*Theology is a genesis*" and secondly, that *arcology stands for a truly human* landscape that fosters such genesis.[22]

The contemporary environmental crisis really stems from the corrupt values of contemporary industrial civilization, or simplification, giganticism, and ephemerality, which disrupt the spirit and dissipate the matter needed to advance these vast projects.

Attacking the sprawl of contemporary metroplexes with all of their self-destructive tendencies is a spiritual crusade for Soleri to be waged by divinely charged design theory, engineering innovations, and communal revitalization. Sprawl violates the complexity/miniaturization/duration commandments of cosmogenesis; arcology would fulfill them "in such a way as to cause matter to transcend itself."[23] It registers this power because arcologies are engines of "the urban effect," or mediations of "the transformation of mineral matter into mind via the potentially unlimited power of complexification and miniaturization."[24] To invent "the idea-city" of arcology is to reinvent the ordered housing of humanity's mass energy out of the esthetogenesis implied by the urban effect. Thus, along with the urban effect, cosmogenesis becomes being. Human society, when rightly housed and morally aware, will be intent on "the extrusion of meaning, lasting meaning, out of a mineral (mass-energy) reality, powerfully but not fatally swaying the conscience generating within it. . . . the fruition would be the seed of the Cosmos, the Omega Seed, and within it the resurrection/manifestation of all and everything. *That is the end of time and the full dominion of duration.*"[25]

The Arcological Effect

The theological twist in Soleri's thought is directly tied, by virtue of his reflections on the nature of life and consciousness, to ecology and technology. To the extent that human beings make, and are in turn made by, their environments, Soleri proposes that his arcological enterprise can rationalize the processes of their evolution. The ends of complexity, vitality, and miniaturization must be kept in mind as societies design their habitats, because that puts them in a much better position to solve many of today's most pressing environmental problems. Arcology follows these principles strictly and, in so doing, captures all of the beneficial payoffs of six major architectural "effects" for human utilization. Although they are simple, they also are a remarkable solution for many of today's environmental problems:

The *greenhouse effect* is a membrane that seals off an area of ground that can be cultivated, extending the growing season to practically twelve months, and also saves a great amount of water. . . . With the "greenhouse," one has intensive agriculture, limited use of water and extension of seasonal cycles. This is the *horticultural effect*. Then there is the *apse effect*. Some structures can take in the benign radiation of the sun in the winter months, and tend to cut off the harsh radiation of the sun in the summer. By the *chimney effect*, which is connected with the greenhouse effect, one can convey, passively, energy through the movement of air; the heat from one area to another. So we have these four effects; there is also the capacity of masonry to accumulate and store energy—the *heat sink effect*. With relatively large masonry, one can store energy during the warm hours of the day, and give it out during cool or cold hours of the night. The intent is to see if these five effects can be organized around what I call the *urban effect*. The urban effect is the capacity of mineral matter, to become lively, sensitive, responsive, memorizing. . . . If we were to coordinate those six effects together, then we definitely could save on resources like land, water, time, energy, materials, and have a better ecological sanity.[26]

Arcological structures can harness all of these effects together by suspending the two-dimensional explosion of urban areas in a new three-dimensional implosion of urban structures. Radically reconceptualizing the morphology of individual houses, towns, and cities as social aggregates with immanent rational potentialities, instead of seeing them as personal real estate embedded within social irrationalities, is extremely difficult. But, it is possible, and Soleri sees his own arcological experiment, Arcosanti, as "an ongoing process pursuing the urban effect within the context of the other five effects."[27]

The arcology is pure ecological transformation in that it seeks to harmonize cities as "a biomental organism contained in a mineral structure" with their biophysical environments.[28] The scatterization of automobile/railway/streetcar urbanization is attacked by miniaturizing the logistics of energy and matter into superdense structures. Soleri believes that these systems will, in turn, check the degenerative decline of current cities by reducing the waste of energy, material, time, and labor used simply to maintain their irrational and antiecological forms. Arcologies are not utterly improbable utopias. As Soleri observes, the naval architecture of today's passenger ships is very closely related:

> the common characteristics are compactness and definite boundary; the functional fullness of an organism designed for the care of many, if not most, of man's needs; a definite and unmistakable three dimensionality. . . . the liner, the concept of it, can open up and, retaining its organizational suppleness, become truly a "machine for living," that is to say, a physical configuration that makes man physically free.[29]

This analogy is exploited in several of Soleri's proposed arcologies, including Novanoah I (a city for 400,000 to float in coastal waters or on the open sea), Novanoah II (a city for 2,400,000 also to ply some ocean expanse), Noahbabel (a city of 90,000 to be fixed in the sea off of some coastal setting), and Babelnoah (a city of 6,000,000 for a flat coastal region).[30] Of course, all of these imaginary cities, which are little more than illustrative drawings published twenty-five years ago, have never been drawn in any detail or planned on a full scale, although they effectively substantiate Soleri's analogy between large passenger ships and arcological cities.

Once these urban images are made plausible as extrapolations from ocean liners to ocean cities, Soleri articulates other designs for many different land-based sites: Arcoforte (a city of 20,000 on a sea cliff), Babel II A (a city of 800,000 on marshland), Arcvillage I (a farming community of 9,000 near good farmland), Logology (a city of 900,000 for hilly floodplains), Arckibuz (a desert village for 2,500), Arcollective (a cold region village of 2,000), Veladiga (a city in or as a river dam for 15,000), Stonebow (a bridge city of 200,000 to be built over or in a large canyon), and for the ultimate architectural challenge, Asteromo (a city of 70,000 to be built in outer space).[31] Only one of his smallest designs, Arcosanti, has gone from the drafting table to the construction site, but all of its masses are arranged to express the six effects that arcologies must exploit.

Soleri's Contradictions

There are contradictions in Soleri's thinking. Like many teleological theories of history before it, Soleri's project almost presumes the existence of

God cities and God citizens, before they exist, in order to make them come into existence. Surrounded by a culture that obviously accepts, if not openly chooses, to produce urbanization on the order of the obscene metropolitan sprawl found in Atlanta, Dallas, Los Angeles, Houston, or Phoenix, Soleri slogs forward at Arcosanti with a handful of true believers sharing a vision of the ecological good sense in his project. Content with casting wind-bells and evangelizing the few tourists who show up every day for tours of Cosanti and Arcosanti, these faithful few already perhaps live his "esthetogenetic life," transforming brass into bells and concrete into a city. By and large, however, they are young, transient, and enthusiastic residents of a place that lacks any lasting economic or social function beyond demonstrating the possible truth of Soleri's imagination to a civilization that denies it daily in everything that it does. Although the state of Arizona's chambers of commerce are more than willing to promote visits to Arcosanti in hotel lobby flyers and travel magazine write-ups if it brings new visitors to experience the region's many antiecological lifestyles, the state has done little to advance the spread of Soleri's ideas.

The almost unreal waste of everyday life in cities such as Phoenix is a strong boundary condition, constraining efforts to make something like Arcosanti from becoming real. And, with the informational revolution, many even convincingly argue that Soleri's arcologies are absolutely obsolete technologies for advancing the urban effect. Given the new media, such as cable television, mobile personal telephony, global satellite-based communication, and personal computers, two-dimensional scatterized sprawl can emulate three-dimensional complexity—now even on a global level.[32] "Cybertectures" for a third nature are displacing "architectures" of second nature, which might doom Soleri's arcological dreams to sink into the telematic rifts of transnational infostructures. If complexification can advance informatically, then why bother with material manifestations in architectural megastructures? Miniaturization is measured more rightly in bytes and gigaflops, not square meters or population densities. Soleri, of course, protests. In his mind, informationalization makes arcologies even more necessary and desirable. If informational technology is misapplied to the partial rationalization of scatterization, it will only postpone the collapse as it aggravates its consequences. Nonetheless, in some respects, informationalization does begin to sublate the arcological effect, enabling one to treat the entire planet, when it has many mature infostructures up and running at some point in the future, as one gigantic arcological megastructure on the informational plane.

For a totalitarian civilization that mystifies its own totalitarianism, as Marcuse claims, in the ideologies of choice, liberation, and convenience, Soleri's urban experiment can be explained away by neoliberal opponents as a communitarian nightmare: an environmentalist's anthill where no one can have a car, plant a lawn, or build his or her "own" house because the Master Builder as Founding Legislator has deemed them all to be evil. He probably is right, but this sort of right guy has been finishing last in modern industrial cultures for three hundred years.[33] At best, as a real-estate enterprise, Arcosanti may only occupy another small niche in the immense food chain of Arizona's other consumer-segmented lifestyle villages, such as Sun City and Green Valley for oldsters, Scottsdale and Oro Valley for transient yuppies, or Ahwatukee and Chandler for young families. For hard-core ecological dreamers at the tail end of the market's psychodemographic distribution, who might otherwise settle for a New Age 1960s aura in Bisbee or Jerome, Arizona also has Arcosanti. But, as real estate, Arcosanti also is utterly illogical in the current capitalist space economy of place valorization and devalorization. As a devotion to permanence, no current business categories can deal with it. Its projected horizon for economic life is completely off the ten- to fifty-year scale of amortization used in Arizona's mortgage houses. Compared to a city like Phoenix, where blocks and blocks of dream houses built in the late 1970s already are turning into slums, Arcosanti is totally irrational. Arizona's entire political economy is based on rapid cycles of growth and decay in its real-estate markets, whereas Arcosanti is predicated on ripening slowly as a slow-growth, or even a no-growth, society.

In some ways, a Phoenix or a Los Angeles—in their full-blown 1945–90 Cold War forms—perfectly represent the urban form in an era of thermonuclear confrontation. Spreading out from the ground zero of downtown government centers or central business districts, which lack any authentic "downtown" or "uptown" urban vitality, these urbanized conglomerations are thin accretions of impermanent structures operating under the menacing shadows of intense thermonuclear explosions. Lacking any civic center holding them together, such places have gained their social identity mainly in and from televisual images. Nightly news broadcasts, detailing the murders, ball games, and rainfall occurring within their borders, create a sense of "Phoenix" or "L.A.," but, for the most part, there still is "no there, there." Built around the high-technology industries used to construct the thermonuclear delivery systems of the federal state, Cold War anomie and superpower angst continuously deconstructed their community, reducing them to settled assemblies of megadeaths awaiting Soviet megatonnage that in the meantime also provide the grim televisual settings for police-state morality plays such as *Dragnet, CHIPS, Baywatch,* or *Cops.* In a sense, they are cities where the bomb already has dropped, leaving a uniformly drab collection of rubble where atomized communities of isolated people scurry from flimsy shelters to clogged streets to dilapidated commercial strips to abusive workplaces stretching for hundreds and hundreds of square miles.[34] Only an H-bomb could destroy them, but perhaps only an H-bomb-centered political economy likewise could create such "urbanized" places without leaving any semblance of a city within their "city limits."

At the same time, Arcosanti now is languishing, because it does need a political economy to operate: an entire industry to sustain it, a real community to build it, a living society to maintain it by putting all of its energies into its realization. Although Soleri argues that a city is essentially an informational construct, Arcosanti may prove how limiting a purely informational economy can be, particularly if it has only two product lines: wind-bells and tourism. Now, at best, it is a prophet's manor or a monastic order's abbey, but not a city.[35] Lacking more diverse ties to other industries, most markets, or more information, it cannot really become a city. It only is a house of ideas, and it may well become a ruin once the idea man is gone. Something rising on the scale of Arcosanti really requires a corporate or state agency to raise the capital, labor, and materials to make it operate, but under current conditions neither Arizona nor Soleri will allow this to happen. Thus, Soleri's city is simply a mineral structure, lacking the biomental organism to sustain it.

An intentional community of fifty or sixty has been created over twenty-five years to occupy it, but its spaces were planned to hold five thousand to six thousand people. To the extent that it only has attracted 1 percent of its ideal population, Arcosanti probably is a failure even as a monastic community.

Soleri's cities have been maligned for their anthill qualities, but he actually has changed only a couple of design assumptions in urban engineering: first, moving from automobile to pedestrian transport and, second, shifting from many ministructures to constructing a few megastructures. His changes, however, may not radically alter the city's ecological imprint on the countryside. A million people in an arcology might not be more ecologically viable, because a modern city is merely the material manifestation of many different global flows of people, energy, and products intersecting at one particular site. Soleri perhaps prematurely attributes many environmental problems to the shape of the city's architectural nexus with Nature—or the city as material structure/real estate/geographical site. Much could change, and undoubtedly even be improved, in arcologies of his design, but he is naive in believing that what exists now as metropolitan sprawl is not also an arcology—only of a highly deviant type. Destructive global flows of capital, labor, energy, material, and ideas might just as well, or even more easily, circulate through his arcological structures, altering only a few local ecologies, and not improve global environmental imbalances all that much.[36]

Soleri's arcological designs also fly in the face of many contemporary environmentalists' notions of living in harmony with Nature. He openly admits that nine or ten billion people, or double the planet's current population, could inhabit the earth, if only its urban infrastructure could be arcologized to fit his specifications.[37] This is not the picture of "sustainable ecological development" envisioned by most of today's Greens. With this assertion, he also remains very eclectic about choosing his proposed sites of construction. Building tremendous megastructures over undeveloped canyons, in river courses, or along cliff regions is not the aesthetic of frugality and appropriateness followed by the followers of deep ecology, ecofeminism, or even social ecology. The public benefits of following Soleri's path to ecological frugality could prove very satisfying even to some radical ecologists, but his methods have the signs of inappropriate technology written all over them inasmuch as he calmly plans for a doubling in the earth's population.

Arcosanti is not a complete arcology; it is at best a fragment of a future yet to come. In fact, many Arcosantis, as greenplexes for ecologically inclined consumers, could be built within the current frameworks of advanced industrial civilization without even rippling many of its antiecological qualities. Only if scores of arcologies on the scale of Novanoah II (2,400,000 inhabitants), Babelnoah (population 6,000,000), or Logology (900,000 citizens) could be built all over the planet would Soleri's arcology be a working idea. By remaining trapped in the spiritualistic echo chambers of 1960s consciousness-raising, Arcosanti has failed to anticipate many of the most significant global changes happening in the 1990s. Its New Age auras have totally eclipsed its inhabitants' abilities to adapt Arcosanti's ecological lessons to the new times of post-Fordist informationalism.

The political economy of Arcosanti today, then, is not tremendously political, nor is it remarkably economic. Instead of building a city for a society with many of its citizens already in place, this "urban laboratory" has become more of a rural retreat from the Sun Belt frenzy of Phoenix for Soleri and his coworkers. Sitting so close to I-17, a major north-south route from Phoenix to the Grand Canyon, Arcosanti increasingly is operating daily as just another roadside attraction, starring Soleri and his followers as funky inhabitants of a bizarre living history village from the 1960s. Whereas Williamsburg features fragments of the past dressed up by the present as American history, Arcosanti presents fragments of a future that gets dressed down by curious tourists from the present as a fascinating but failed expression from the Age of Aquarius. For twenty-five years, twenty-something dropouts have been casting wind-bells, baking organic muffins, leading instructional tours, or raising concrete walls at Arcosanti. Instead of creating his city, Soleri has established a theme park. Like Williamsburg's tasteful simulation of English colonial life under the Hanoverian kings, Arcosanti too is an artful simulation, only it remains stuck on memorializing what might have been had some elements of the Woodstock Nation assumed power. It is a nice place to do workshops, as the frequent customers in its Elderhostel, silt-casting seminar, and New Age concert businesses prove, but most people would not want to live there. Thus, it serves unwittingly as a vivid affirmation of everything that modern consumerism associates with the "small is beautiful" and "voluntary simplicity" philosophies of the 1960s—living at the dead end of a dirt road in the desert amid the ruins of a commune to handcraft wind-bells and print manifestos to sell to carloads of Japanese tourists.

Not surprisingly, the capital accumulation prospects of this quasi-manorial economy have been almost nil. Arcosanti's annual construction budget is about two hundred thousand dollars, or much less than what many Phoenix suburbanites shell out for one single-family house down the road from Cosanti in Paradise Valley. Arizona's population almost doubled from 1970 to 1995, but Arcosanti's fifty or sixty permanent residents are a barely stable settlement. Moreover, the ecological credo of Arcosanti is the cause of this steady state in its economy. Seventy miles south, Phoenix is a growth machine that mushroomed from 150,000 people in 1945 to 3.5 million in 1995 by treating both architecture and environment as disposable personal commodities. Housing is the engine of its growth machine, because it is a cheap, mass-market commodity tossed together in sixty days out of chipboard, two-by-fours, and spray-on stucco on tiny lots jammed together near a golf course, playground, or shopping mall. People live in it for five or six years, then resell to move up and farther out. Sold and resold for fifty or sixty years, it ends its economic life as a freeway teardown, in a barrio, or simply abandoned in the city's spreading urban core. Huge amounts of water, electricity, and material resources are needed to keep this home habitable in Phoenix, all of which come from outside of the Valley of the Sun at unsustainable levels and rates. Still, there has been constant growth for five decades by reproducing the same cycles over and over. This new "sixth C," or construction, is what keeps the Arizona lifestyle rolling.[38] Without it, the state again would become the backwater that it was before this political economy was invented in the 1940s—underpopulated, economically stagnant, and politically insignificant—as it pursued the traditional "Five Cs" of citrus, cotton, cattle, climate, and copper in a semi-Third World extractive economy.

If this growth machine is Arizona or America, then, Arcosanti is "un-Arizonan" and "anti-American" through and through. Something that lasts for generations, some place that is ecologically sustainable, some community that is rooted in the environment is not part of the contemporary lifestyle in Arizona or America. Soleri cannot succeed, because everything that he regards as ecological failure in Arizona is also the sine qua non of its irrational but materially quite real economic prosperity. It is this stark juxtaposition of Phoenix with Arcosanti that must be studied for its quintessential confrontation of antiecology with ecology. From the antagonism of these competing urban forms, much can be learned.

Summary

At the end of the day, Soleri and Arcosanti represent a contradictory proposition. Various ideologies and utopias freely intermingle in everything he has achieved. Soleri's utopian arcologies at first glance often do seem to be the wish dreams of a civilization yet to be realized; however, after a second look, they also can be seen as ideological fictions for stabilizing many myths in the present social order. Indeed, these ideological qualities are the most vexing product of his work.

First and foremost, Arcosanti might be regarded as just another totalitarian technofix for the problems plaguing contemporary industrial civilization. Even though it is dressed up in the clothes of theological celebration, a Soleri arcology is a tremendous megamachine dedicated to rationalizing the metabolism of urban economies and societies by building bigger to last longer. Each arcology rests on a design for concentrating the production/consumption/circulation/administration of commodities in the bowels of an immense megastructure, while arraying the spaces for residences, schools, entertainments, and the arts on its top or along the exteriors, like some superluxury steamer permanently run aground on the earth. Thus, once again, rational design and logical construction on a gigantic scale by a world-historical genius are represented as the only true path for the ecological salvation of society. Ironically, this project probably has been stunted by being so closely held; it really needs opening up to popular feedback and community input to actually succeed economically or socially.

Second, Arcosanti comes off as an architectural concretization of the vanguard schema of enlightened rule. Left to their own devices, the citizens of Arizona, America, or the world apparently will accept a Phoenix, Los Angeles, or Mexico City as their habitat. Regrettably, such urban forms are an ugly urban sprawl, even though their inhabitants can and do deal with the disorder. To survive, Soleri believes that this civilization needs new leadership: rational, enlightened, arcological guidance from above. Missing a sacred element in its secular hustle for progress and fun, this civilization also could use a higher purpose, such as enacting the cosmogenetic realization of some God-force in human history. Soleri's arcology occasionally smacks of a puritan desert despotism, albeit virtuous and enlightened, not terribly unlike those that have already crashed and burned elsewhere during the twentieth century. The architect, as master builder, will dictate the shape and function of space down to the one thousand to two thousand square feet each family will receive.[39] In these small cabins, and only there, like passengers on a ship or slaves on the plantation, might personal desires be accommodated and expressed. The thorny questions of real property law, private ownership of wealth, and public corporate organization, which are the material forces that actually animate any contemporary city, are ignored in this totalizing vision of ecological order. And, inasmuch as legal and economic issues are neglected in Soleri's grand architectural design, one sees a major reason for Arcosanti's pathetic record as a working human settlement.

Third, Arcosanti could be dismissed as a true utopia. That is, it is "nonplace," because it is in a no place. Plunked down in the middle of nowhere with no industries, little arable land, few transportation links, minimal scenic attractions, and hardly any people, why would anyone want to live there? Everything about it, even if it were completed, signals that it is the negative antithesis of all that is, rightly or wrongly, regarded as life-affirming in the scatterized metroplex. Cosmogenesis in the Arizona desert will lose out every time when put up against plain old technogenesis in Houston, plutogenesis in New York City, or infogenesis in San Francisco. If Arcosanti did not exist, the average American real-estate developer could argue that it would be necessary to invent it to reaffirm the virtues of the vices Soleri decries in large contemporary cities like Phoenix and Los Angeles. The multiple myths of mobility, fun, pleasure, power, and convenience that contemporary consumerism thrives on may well be false promises, especially when viewed from an ecological perspective. Yet they are believed to be true, so they become true. Arcosanti's misfortune, ironically, is to serve as tangible proof of a utopian alternative's undesirability by showing itself as an ideology based on what today's average suburbanites fear would be a lifestyle grounded on excessive immobility, labor, pain, powerlessness, and inconvenience.

The complexity and energy of the urban effect is real. Glimpses of it can be gained from midtown Manhattan, Chicago's Michigan Avenue, downtown Boston, or San Francisco's Market Street. However, most of these energies are generated by major corporations, large banks, and big governments. The arcological challenge is how to capture and concentrate these effects in more popular forms of municipal commonwealth or urban communitarianism without succumbing to Soleri's utopian ecologism. Soleri professes to have no preconceived plans about how to organize the civic administration of his arcologies. Instead, he believes that they are like pianos on which almost any kind of music could be played by the societies that acquire them. This metaphor is alluring, but it overlooks how directly any arcology, like a piano, must express the collective values and personal tastes of those peoples who invent and build it. It takes a complex society to build a piano, and only particular types of social individuals can, in turn, play highly complicated and varied kinds of music on this instrument. Arcosanti has failed as a city thus far by not accounting for this essential civil element in its designs; not only must the piano be built, but so too must players be trained, listeners educated, concertos composed, concert halls designed, and piano factories constructed for "piano playing" to succeed.

Fourth, in casting arcological invention as an antiurban return to Nature, Soleri picks the fight (out of some sense of purity) that cannot be won and passes on the battles that must be won (on the basis of practicality) in the existing city. By presenting the arcological alternative as one more futurology for tomorrow by setting his new megastructures into fantastic unspoiled bioregions, he ignores the possibilities for materially improving all those existing arcologies that already are failing their inhabitants now on-site in large cities. Instead of reimagining urban infrastructures from the ground up on new unspoiled sites, why not push people toward thinking about arcologically retrofitting existing cities? Phoenix easily could become a series of smaller, interconnected arcological structures, allowing the desert, the river bottoms, or even the citrus groves it has annihilated with unchecked suburbia over the past five decades to return. Rather than avoiding real battles over the prevailing approaches to land use, shelter construction, property creation, habitat destruction, and urban culture by retreating to a utopia in the desert, why not join in the political struggle to redefine everyday life by pushing all existing scatterized arcologies to shed much of their irrationality in becoming more like Soleri's idealized concentrated arcologies? Over the past twenty-five years, huge malls and commercial projects occupying much more land than many of Soleri's designs have been built very rapidly in Phoenix. Indeed, large tracts of land from federal reassignments of property, failed commercial developments, or dead-and-gone 1950s shopping malls have become available for arcological experimentation in Phoenix, but these opportunities for real transformative experimentation with the structures of a thoroughly antiecological city have been lost for the lack of a Solerian arcological will, imagination, or desire.

Fifth, Soleri's fairly impoverished categories of critique chalk up the two-dimensional sprawl of a Phoenix, Los Angeles, or Houston to the effects of one technology—the automobile. On one level, this analysis is correct, but remarkably incomplete. In fact, one might argue more effectively that these sorts of cities grew hand in hand with the development of the interventionist federal state during the Cold War era. The economy of these Sun Belt urban sprawls was deeply embedded in the high-technology astronautical, aircraft, and electronics industries that rested at the heart of the Fordist American defense economy. Such cities also were made possible by federal subsidies to interstate highway construction for their transportation grids, hydroelectric dam construction for their electricity and water supplies, cheap residential construction for their housing stock, and corporate agribusiness construction for their food and fiber needs. The automobile, air conditioning, and the mass media provided the requisite technologies to move people and goods, maintain climate-controlled habitats, and organize thoughts and desires of millions of people moving into such cities from all over the nation almost overnight to live and work in an entirely new form of urbanized space. Yet, like the national regime that made these urbanisms possible, these urbanized concentrations of population largely lack any real organic community, localistic particularity, or democratic grounding. Such sprawls are simply temporary housings or transitory habitats for thousands of families moving from place to place by corporate directive or governmental design to sustain the federally funded growth machines of the Cold War economy. Now, as the military goods once produced in helicopter plants, aircraft factories, and missile installations in Phoenix, Los Angeles, and Houston are no longer in demand, and the federal largesse that built the dams, highways, and suburbs that made such places possible is drying up, these two-dimensional cities will face severe new challenges simply to survive.

The concentration of population and production in compact megastructures, as Soleri proposes, might well be a highly rational response to the post-Cold War political economy of flexible accumulation. Cities no longer can necessarily count on major subsidies from higher national authorities to pay for inefficient transportation systems, inhospitable housing tracts, and expensive water-supply systems. To respond to the demands of global rather than national modes of high-technology production, cities will need compact, flexible, inexpensive spaces for offices and factories close to their population centers. Reducing everyday frictions caused by the urban sprawl, traffic gridlock, and social decay embedded in the automobile-based Cold War city's civil engineering systems and architecture will require radical innovations. Building variants of Soleri's arcological settlements might be seen as taking decisive steps toward adapting to the transnational economies of post-Fordism.

Yet, the isolated personalities and atomized families nesting in Cold War-era suburbias rarely foster the sort of social individuality needed to commit any individual's life or every family's fortunes to the permanent civic commitment of constructing an arcology. Instead, one tends to find many more frequently underemployed, unstable drifters, such as those recruited into the "Viper Militia" of Phoenix, who allegedly plotted to bomb the visible centers of federal authority in the Valley of the Sun's built environment to launch a popular uprising against the regime that nurtures such cities.[40] As even the Viper Militia members grimly perceive, to work well for all, these existing arcologies really would require a new form of state, rooted in the practices of popular communal governance. Such populist commonwealths might channel Soleri's "urban effect" toward generating more rational relations of production and consumption better suited to constructing an arcological city. And, in turn, the urban effects of these arcologies could charge new populist commonwealths with tremendous vitality and creativity.[41]

Despite these reservations, the positive utopian energies of Soleri's work should be both acknowledged and affirmed. He must be praised for beginning to rethink the dynamics of human ecology. Apart from his odd Teilhardian foundations, which may or may not have merit for many readers at this juncture, he invites us to reconceptualize how extensive the human impact on the environment already has been. Arcologies already exist, but they work in scatterized, megalopolistic, and inefficient forms of transnational corporate commerce, wasting energy, material, and lives in an ongoing degradation of all human and non-human life on the planet. The environment is not something outside of these transnationalized high-technology arcologies; these arcologies have, in fact, iterated themselves out onto a global scale of operation, making their megamachineries "the environment" for most humans and many species of planets and animals. Ecological change must, as Soleri argues, renaturalize these environments by promoting the benefits of complexity and miniaturization.

There are workable ideas for communities of two thousand, twenty thousand, or even two hundred thousand inhabitants in Soleri's visionary designs. The first reaction of many to his ideas is disbelief: "it can't happen here," "too big, it won't work," "nothing like this can exist." But, some stand-alone megastructures, such as the Pentagon, the World Trade Center, the Omni in Atlanta, or the many "Galleria-type" megamalls rising all across North America, have been built by this society, even if the existing corporate relations of production pervert them to fulfill essentially antiecological uses. New popular movements and/or existing urban communities could choose to follow arcological ideas, as Soleri has been testing them at Arcosanti, in new attempts to build a truly more ecological society. The everyday frugality that many ecologists claim is essential for a sustainable society waits to be realized within the frameworks provided by Soleri's arcological ideas. In the post-Cold War world, which apparently needs some new transcendent purpose for maintaining any focused collective action, building arcological cities could provide larger goals for successive generations of human beings to devote their energies and labors to good uses in a particular ecological setting. But these efforts also must move beyond Arcosanti, mobilizing arcological changes at home, in the neighborhood, and downtown all over the nation in order to maximize their ecological impact.

9

Community and Ecology: Bookchin on the Politics of Ecocommunities and Ecotechnology

This chapter reexamines some possibilities promised by a political economy grounded on the principles of social ecology, as they have been elaborated by Murray Bookchin. It considers some alternatives for populistic/localistic economic exchange, political organization, and technological production that Bookchin sees as real opportunities for those following the guidance of social ecological discourse. Bookchin's use of "social ecology" describes his application of ecological reasoning as critical social theory to questions of radical social, political, and economic change. In defining social ecology, Bookchin argues:

> Our own era needs a more sweeping and insightful body of knowledge—scientific as well as social—to deal with our problems. . . . We must seek the foundations for a more reconstructive approach to the grave problems posed by the apparent "contradictions" between nature and society. We can no longer afford to remain captives to the tendency of the more traditional sciences to dissect phenomena and examine their fragments. We must combine them, relate them, and see them in their totality as well as their specificity. In response to these needs, we have formulated a discipline unique to our age: *social ecology.*[1]

Bookchin's ecological critique is not flawless. First, his visions of an ideal democratic city in highly romanticized Hellenic terms often seem unworkable, infeasible, or unrealistic for addressing the scale of the many problems facing contemporary urban centers. The techniques for getting from the prevailing forms of overdeveloped urban life to his ulti-

177

mate state of municipal confederalism plainly are often more suggestive than definitive. Second, his somewhat uncritical notions of "history," as some sort of univocal, true voice of substantive reason and communal identity, frequently sound like discursive writs of ecoactivist empowerment. Any sense of historical narrative's fictive, mythic, or delusional qualities is essentially ignored rather than acknowledged as a possible pitfall as Bookchin's writings reference historical treatises as if they were an objective record to divide what can be from what has been. Third, Bookchin essentially sees humanity as the consciousness of a purposive and ordered Nature. This position legitimates much of his ecologically directed project for social transformation; but the basis of this initial position, in which Nature has some apparent entelechy that humanity must not alter by either any ill-considered technoeconomic project or the sin of prideful domination, basically is articulated more often than not as a stance of faith rather than as a rigorous rational argument. Such ontological/epistemological positions are defensible, but he tends to not systematically defend them. Consequently, his Nature philosophy often does seem strikingly atavistic in an era of post-Fordist informationalism where "Nature" becomes, as chapters 4 or 5 suggested, an artificial simulation or engineering emulation rather than remaining something like a cosmic fact. Nonetheless, his articulation of social ecology as a critical theory—despite these considerable flaws—remains one of the most vital projects on the scene today.

In one form or another, the essential spirit of social ecology can be traced back to Rousseau, Blake, Morris, or Kropotkin. Bookchin, who is its most articulate contemporary exponent in North America, has engaged in systematic political debate in favor of social ecology since the late 1950s, mainly in response to the destructive environmental changes being wrought by the corporate economy soon after World War II. As Bookchin suggests, an entirely new mode of highly destructive industrial production, based mainly on the growth imperatives of transnational corporate capital, successfully has penetrated virtually every traditional economy, local community, and bioregional ecology on the planet. Although such economic practices increasingly are characteristic of all advanced industrial societies—whether they are corporate capitalist, state socialist, or national corporatist in their predominant organizational and ownership patterns—the most extensive elaboration of its adverse ecological effects can be found in the corporate capitalist economies of Western Europe, North America, and Japan. The political economy in Bookchin's social ecology aims a comprehensive critique at these imperatives—endless growth, unregulated waste, overspending of energy, overproduction of useless things, technical expertise over users' needs, productive labor displaced by underproductive leisure, and a consumer society imposing itself on a conserver society.

The Roots of the Current Crisis

Ecological cycles of natural environmental reproduction are global, borderless, and transnational. With the waning of the Cold War, the allegedly ever-present threat of communist subversion, which once justified and guided globalist, cross-border, transnationalized state and corporate projects of modernization, is lessening, if not entirely disappearing. Yet, in the aftermath of the Cold War, it is now easy to see that the global economy that was constructed during the course of anticommunist interventionism by loose alliances of the professional managerial classes in academic, corporate, financial, or government organizations remains in place and growing. Its very functioning, in turn, continues to ravage the world's environments at dangerously high rates of local destruction that have global implications. Many of these "new class" worldwatchers are, to put a new twist on a popular green slogan, very cynically "thinking globally" and opportunistically "acting locally" to shore up their own political and economic authority as symbolic analysts in control of today's fast capitalism. By defending the environment, containing waste, and watching the world in ways that suit their use of power, their clients and consumers, only now on a global scale, will remain passive, dependent, and powerless.

"Thinking globally" by such new class agents often articulates the environmentality of imperializing blocs of growth-minded nation-states and transnational corporations. It demands submission to formal codes of instrumental rationality capable of generating images of reality and then judging in terms of elaborate statistical models, which are, in turn, assumed to be identical to, and exhaustive of, the substantive social, economic, or cultural forms of life captured within their disciplinary grids.

To Bookchin, such reasoning can only be reductionistic, instrumentalist, and destructive. Everything not disclosed by its professional decision-making rules of statistical standard deviation or mathematical multiple regression is crushed, ignored, or distorted to fit the uniform results standardized on the expected slopes of mathematical prediction.

The new class empowers itself and disempowers immediate local communities all over the globe through a simple bargain. To accept them and their systems of knowledge and power, the short-term flows of goods and services on a global scale will continue, and might even grow for a few more decades. To resist them, the flows of goods and services may slow or stop almost immediately, which could lead to a major social and economic crisis. On the apparent virtues of this trade-off, the new class lives for and within the structures and codes of its knowledge and power. No longer local, their most basic community is clearly trans-national and highly homogenized. The elite professional spaces delimited within telecommunication links, jetports, high-tech office complexes, upper-income neighborhoods, new science centers, and powerful bu-reaucratic agencies that they work within all tend to reproduce the same sets of expectations and practices everywhere in the transnational net-work. Community for them often is national or transnational, not local or regional. Rather than living in and by one ecoregional setting, they live within the mass mediascapes and commodified cyberterrains of these global networks. "Acting locally" by the new class, while "thinking glob-ally," simply means surfing in Maui and thinking about the next busi-ness deal in Manila, working at home but commuting cybernetically to the bond business pits in Tokyo, or contributing to the local PBS tele-vision station to pay its share of producing a documentary in London about saving elephants in Africa.

Beyond the green rhetoric of new class power, however, these in-creasingly borderless minimal "communities" also are ensnared in global networks of exchange, living off of millennia of slowly accumulated fos-sil fuels. Maximizing mobility for people, goods, services, and technol-ogy rather than ensuring sustainability is their major operative logic. Often none of their vital ecological supports are in the immediate envi-ronmental vicinity. Environmental inputs are not used on a sustainable scale appropriate to each bioregional setting. Instead, these supporting flows are sourced from around the larger nation-state or even the world, bound together by the wasteful expenditure of scarce nonrenewable energy. And, because of their location at core operational nodes in the global network of such transnational flows, some neighborhoods, cities, regions, and nation-states also either enjoy tremendously unequal rates of excess consumption or suffer from outrageously inadequate levels of basic goods and services.

The new class is, first and foremost, globally flow-minded, or grounded in transnational exchange, and not locally place-minded, or embedded in a particular ecoregional setting. Living lives of mobility, such symbolic analysts often treat most natural places frivolously or disrespectfully. Any place is usually no better than any other; indeed, every place is equally subject to potential destruction as a development site, economic raw material, or fresh market to sustain the flow. Building community around a place, accepting it on its own ecological terms, working to adapt a sustainable way of life in it, and cherishing it for its unique differences are notions foreign to the operational logics of new class power and knowledge. Living for the flow, and sustaining themselves in various impermanent eddies in its streams, new class power follows larger national and transnational agendas, which ultimately are ecologically unsustainable.

By "outsourcing" the material basis of our lives, or organizing everyday life around importing incomplete segments or partial components of the total array of goods and services needed to survive, the new class state and corporate planners have empowered themselves by breaking many communities' communal ties with each other as well as most of their organic ties to their immediate ecoregional settings. And, by moving communities toward "specializing," or exporting their own small, limited stream of componentialized goods and services in sufficient volume to run a basically balanced account of trade with other communities caught all over the world in the same networks of exchange, most individual localities either prosper or fail in global zones of economic development managed by new class experts in state and corporate bureaucracies. Perhaps profitable in a purely exchange-based system of accounts drawn up for the quarterly reports to new class management, these cycles of global trade are, in fact, extremely unsustainable. A thorough ecological audit of this disastrous model of growth suggests that new class "worldwatching" in the 1990s is not much more effective than Reaganite bank regulation during the 1980s.

The populist-communitarian critique of the new class and its liberal agendas for social administration usually takes one of two lines of attack in assaulting the theory and practice of new class bureaucratic domination in the United States today. According to Walzer, the first communitarian response to liberal systems of administration, such as Alasdair MacIntyre in *After Virtue*, "holds that liberal political theory accurately represents liberal social practice."[2] Thus, for populists to criticize liberal society, all one has to do is accept liberal theory seriously on its own terms. In other words, the vision of subjectivity propounded by Locke, Smith, or Bentham is true. Hence, this critique argues that contemporary society is becoming full of rational economic agents, existentially atomized, isolated, and fragmented in an endless game of complex rationalistic calculation. But, as Lasch observes, most ordinary people actually do not want to live this way, and they basically associate it with new class styles of behavior.[3] Left free to choose or granted the right to choose, according to this populist critique, there can be no serious moral criteria guiding individual choice, save personal utility or the unstable interests of loosely aggregated voluntary associations, which inevitably leads to constant crises that only popular-communitarian resistance can solve.

The second populist-communitarian response to new class liberal power, as in Bellah et al. in *Habits of the Heart*, takes the opposite approach as it "holds that liberal theory radically misrepresents real life."[4] In this account, populists suggest that because we actually live in a rich world of complex social ties and intricate communal connections, liberal practice and theory cannot be taken seriously. Rational economic agents, set free to act on their individual utility calculations, have little or nothing to do with actual reality; they are instead the ideological aspirations of those upwardly mobile individuals, like the new class, who leave local communities rather than those who form or remain with them. Again, this seems to be Lasch's point in contrasting populist criticisms of society from below with new class notions of professional symbolic/analytic power from above.[5] Therefore, to criticize the liberal forms of modern society created by the new class, all that anyone must do is reject liberal theory completely on its own false terms, while attending to its actual surviving preliberal, nonliberal, or postliberal attributes that make it work, since they are the only possible basis for building a better popular communitarian order.

Both of these attacks on liberal new class thinking and practice carry an element of truth, but neither one of them goes far enough. From the outset, any political discussion of today's crises must return to the volatile issues of ecology. Lasch raises many critical points about populism as an alternative basis for political organization, but he consistently ignores the traditions in American thought and politics that often have approximated what he hopes to realize, namely, ecological thinking and environmental activism with all their many flaws. Although Lasch does not acknowledge his significance, the most dogged and important defender in the United States of the virtues of popular democratic ecological community since the 1950s has been Murray Bookchin.

Community and Ecology: Localism versus the New Class

Community, as an organizing concept, can and does have many meanings. However, in the frameworks of contemporary new class administrative control, the urbanized community tends to decay into the most "minimal" or very "thin" attributes of communal unity with very little popular form or independent social content. Bookchin observes that

> our urban belts and conurbations are vast engines for operating huge corporate enterprises, industrial networks, distribution systems, and administrative mechanisms. Their facilities, like their towering buildings, stretch almost endlessly over the landscape until they begin to lack all definition and centrality. It is difficult to root them in a temple, palace, public square, or the small, intimate marketplace of craftsfolk and merchants. To say that they have any specific center that gives them civic identity is often so ill-fitting as to be absurd, even if one allows for centers that linger on from past eras when cities were still clearly delineable areas for human association.[6]

As clients and consumers, communities of human beings today in many areas of the United States are not much more than an interacting population of different individuals at a given location (like the suburban households in tract housing developments or the patrons and salespeople at the same local shopping centers) or a stable population of businesses and households organized into discrete geographic-legal units (like the circulating sets of firms and families sited temporarily or permanently in some city, county, or town).

Given the tremendous personal, economic, geographic, and social mobility of contemporary American society, community acquires this thinness because workplace and residence, production and consumption, identity and interests, administration and allocation are so divided in the corporate new class design for advanced industrial society. For Bookchin, these depoliticizing divisions between new class corporate management and the typical suburban consumer result from

> the massive institutional, technological, and social changes that eventually dispossessed the citizen of his or her place in the city's decision-making processes. Urbanization, in effect, both presupposes and later promotes the reduction of the citizen to a "taxpayer," "constituent," or part of an "electorate.". . . Like the modern market, which has invaded every sphere of personal life, we shall find that urbanization has swept before it all the civic as well as agrarian institutions that provided even a modicum of autonomy to the individual.[7]

The division of interests, loss of common historical consciousness, weakening of shared beliefs, and lessening of ecological responsibility in these minimal or thin communities, in turn, are what turns attention to alternative approaches for understanding community.

After Rousseau, Smith, and Marx, conventional canons of social theory almost always have sought to position their understanding of community in discourses of modernization. As Ferdinand Toennies maintains, in the rise and spread of national and international networks of capitalist exchange, the close ties of organic local community, or gemeinschaft, allegedly crumbled into rubble as the formalized relations of contractual national society, or gesellschaft, buried it under the much more formal, organized and rational practices of modernity.[8] "Community," then, is the warm, good, close, past-present conceptual pole of an oversimplified dichotomy opposing the cold, bad, distant, present-future pole of "society." Using this framework, however, any society can be characterized and classified by placing it somewhere between these two theoretical oppositions, although other formal discussions have tried to elaborate on these general distinctions by looking at other similar indicators.

These parallel efforts include Maine (status and contract), Durkheim (mechanical and organic solidarity), Weber (substantive and instrumental rationality), Cooley (primary and secondary groups), Linton (ascribed and achieved traits), and Parsons (systemic or patterned variables).[9] In all of these analytics, however, the primordial sentiments of face-to-face community, coupled with its wholly "premodern" cultural, political, and social institutions under direct popular-communal control, typically have been cast by sociological-political analysis as the stuff of unwanted primitivism, which must be eliminated by the rational progress made possible by elaborate national state bureaucracies and intricate global markets in "modernity." On this point, Bookchin notes how the members of most communities are

> employed in urban jobs, be they professional, managerial, service-oriented, craft-oriented, or the like. They live highly paced and culturally urbane lifeways that lock into mechanically fixed time slots—notably, the "nine-to-five" pattern—rather than follow agrarian cycles guided by seasonal change and dawn-to-dusk personal rhythms. Urban environments are highly synthetic rather than natural. Food is normally bought rather than dispersed. Personal life is not open to the considerable public scrutiny we find in small towns or rooted in the strong kinship systems we find in the country. Urban culture is produced, packaged, and marketed as a segment of the city dweller's leisure time, not infused into the totality of daily life and hallowed by tradition as it is in the agrarian world.[10]

The alleged stasis of such premodern communities provides a legitimating writ for imposing far more fluid, mobile, and variable forms of everyday life, such as those produced and managed by the new class experts, in the diverse voluntary associations of modern contractual society. Instead of traditional solidarity, organic identity, and communal purpose, the dynamism of modern societies brings the possessive individualism, situational rationality, and instrumental self-interest favored by new class bureaucrats and managers.

Once community is framed by these discursive categories, its meaning becomes even more problematic because new class analytic assumptions guard against "thicker" or "maximal" understandings of it, such as Bookchin's notions of popular confederal municipalism, which typically have populist undertones. In fact, modern canons of social theory usually can cope conceptually with these alternative visions of community only in one of three equally distorted forms. The first set of reactions to maximal community usually interprets it as a practical commitment to *conservatism*, or an intense defense by landed or privileged interests to defend undying traditional practices. Yet, in reacting this way, the dichotomous irreversibility of the gemeinschaft-gesellschaft conception of community is, in fact, partially broken, because modern social theory, as articulated in new class social science, must admit that some community-tradition-folk practices do survive and coexist with society-modernity-secular practices. However, the new class stigmatizes a concern for surviving traditions as being conservative, because accepting traditional continuities also implies an endorsement of organic hierarchies rooted in age-old prejudices, inequality, and oppression. As the architect and guardian of modern thin communities, the new class takes great pains to argue that traditional thick communities necessarily would lead to resurrecting these premodern evils in unadulterated form.

If this defense fails, then the second approach to the model of maximal community typically casts it as an unexamined reversion to *romanticism*, or an intense wish to go back to now long-lost, but still not forgotten, sets of primitive simplicities. However, this reaction also undercuts the gemeinschaft-gesellschaft vision of community by acknowledging the moral emptiness of modern society and admitting how the humane closeness of traditional community returns as an unfulfilled utopian image totally opposed to the empty formalistic superiority of modernity. The new class ardently sanctions any engagement with such "voluntary simplicity" or "beautiful smallness" as a very dangerous romantic nostalgia. Rather than embracing the unsettling uncertainty of capital exchange, romanticism would turn the clock back to close face-to-face interaction as a means of recapturing communal solidarity. Yet, as the managers of modern thin community, the new class argues that such forms of communal life never really existed anyway (or they simply are fictional ideal types constructed to make theoretical distinctions), and, even if they did exist, it would be impossible and undesirable to revitalize such maximal communities today (or their actual re-creation would undercut the intrinsic passivity and distance on which new class thin community rests).

The third set of responses to supporters of maximal community presents them as an extremely deviant movement favoring *collectivism*, or an intense impatience with the divisions, instability, and mobility of modern capitalist society that seeks to replace it with some kind of totalitarian collective identity. Still, even this response implicitly challenges the gemeinschaft-gesellschaft notions of community by exposing how incomplete, superficial, and ideological the rational scientific control of modern society actually is today in the hands of the new class. Thus far, the new class, in the United States at least, closely polices these aspirations for maximal community, because some of its radical proponents have arisen in its own ranks and their faith in some yet to be realized future condition of rational totalitarian control can endanger the modest progress being made more slowly and incompletely in the present toward those goals. Instead of dealing with the present problems of minimal community on its own terms, collectivism would rush the clock forward into some technologically defined maximal communitarianism that would fulfill the perfect operational efficiencies of a rational utopia. At some point, the new class may attain these powers, but prematurely calling for their realization in the present only exposes the fragility of their contemporary power and knowledge, as well as sparking fears of the despotic collectivist nightmares of really existing socialism.

Talking about communal life, then, is a hazardous business. None of these alleged conservative, romantic, or collectivist dangers in maximal community necessarily are guaranteed to occur, but the elite discourses dedicated to containing any counterarguments in favor of trying alternative communities continue to issue their ironclad warranties. Unless one accepts the thinnest, most minimal constructions of community, such as those favored by the new class to define geographically or demographically agglomerated, urban groupings of clients and consumers, some bloc of privileged new class interests will feel that its toes are being stepped on. Precapitalist communities, as modern social theory understands them, probably never existed, they would be impossible to revitalize, and their now unknown collective practices cannot heal the individualized injuries that contemporary forms of urban/consumerist community continue to create. As Bookchin suggests, "localism, in fact, has never been so much in the air as it is today—all the more because centralism and corporatism have never seemed more overwhelming than they are today."[11] Hence, the individual and communal desire to reconstitute thin/minimal communities—which are now largely under new class supervision as different populist forms of thicker/maximal community under local popular control—draws stern disciplinary indictments for its localistic or populistic "deviations."

Actually, everyone must realize how many, if not almost all, of the material and symbolic foundations of new class power rest on the results of cold choices made by distant corporate managers and bureaucratic agencies. By interposing specialized technical expertise and complex hierarchical organization, which are centered in massive state bureaucracies and transnational networks of corporate production, between the production of most goods and services and their consumption, the new class disempowers people in their families, neighborhoods, and cities.[12] And, in substituting immense quantities of nonrenewable inorganic energy for more direct and immediate modes of local community production of food, shelter, clothing, energy, and culture, what were once close sensuous ties of human communities to their particular ecological settings have been severed over the past three to five human generations in many regions of the world by these structures of new class power.

The Political Economy of Social Ecology

The political options presented by Bookchin's social ecology require a series of political changes to undo the complexity inherent in corporate capitalist production. Here, then, social ecology asks *the* central political questions: who, whom? Who dominates whom? And, in so doing, who decides whom gets what, when, and how? As Bookchin claims, social ecology does appear to be *the* political framework for engaging in effective contemporary critical analysis inasmuch as "ecology raises the issue that the very notion of man's domination of nature stems from man's domination of man."[13] Social ecology, to a very real extent, looks for the political pivot in all facets of everyday life. The politics of corporate capitalist society, like its ethics, economics, science, and technology, is under the control of professional public administrators, corporate managers, and party/parliamentary politicians. Citizens are reduced to consumers. Yet, merely being a consumer forces citizens "always to accept and take what is and never to share what could be."[14]

The politics of social ecology directly rejects the roles of corporate capitalist society that inform everyone "you are a consumer, somebody *else* is a producer. Production is not a community matter; it is an 'expert' matter, best left to managers and politicians."[15] It rejects this domination because, as Bookchin holds, "the material base for local liberty exists. The decision to have or not have local liberty is just that, a decision, a decision derived from human will."[16] Bookchin maintains that these decisions can be made democratically, effectively, and communally to reclaim personal producership, self-identity, and complete citizenship. Social ecology implicitly asserts that in order to "replace social domination by self-management, a new type of civic self—the free, self-governing citizen—must be restored and gathered into new institutional forms such as popular assemblies to challenge the all-pervasive state apparatus."[17]

To oppose this new class-directed state-corporate regime, Bookchin advocates a politics of *direct action* in which individuals can "function outside it and *directly* enter into social life, pushing aside the prevailing institutions, its bureaucrats, 'experts,' and leaders, and thereby pave the way for *extra-legal, moral,* and *personal* action."[18] In particular, Bookchin favors the confederalist creation of new communities and technologies in the exercise of this direct action. Most important, direct action is the active means for recovering individual control from bureaucratic agencies of the nation-state and corporate firm. In acting directly, Bookchin holds that

> we not only gain a sense that we can control the course of social events again; we recover a new sense of selfhood and personality without which a truly free society, based on self-activity and self-management, is utterly impossible. . . . A truly free society does not deny selfhood but rather supports it, liberates it, and actualizes it in the belief that everyone is competent to manage society, not merely an "elect" of ex-

perts and self-styled men of genius. Direct action is merely the free town meeting writ large. It is the means whereby each individual awakens to the hidden powers within herself and himself, to a new sense of self-confidence and self-competence. . . . It is the means whereby individuals take control of society directly, without "representatives" who tend to usurp not only the power but the very personality of a passive, spectatorial "electorate" who live in the shadows of an "elect."[19]

Direct action constitutes an entirely new political ideal, suffusing all aspects of individual attitudes and behavior. Through direct action, in turn, individuals immediately can begin to reconstitute their own communities and technologies along ecologically more sustainable lines.

The moral necessity of protecting the organic biosphere from advanced industrial societies' tendency to render it inorganic moves social ecologists like Bookchin to call for the re-creation of existing human settlements as "ecocommunities" grounded on "ecotechnologies." Both ecocommunities and ecotechnologies can be characterized by their optimization of personal control, participation, and choice over corporate managers' instrumental rationality. In addition to standing for a totally new spatial arrangement for society, an ecocommunity would become the core of a balanced, humane ecosystem of humans within a carefully guarded organic and inorganic environment. In Bookchin's view,

> if the word "ecocommunity" is to have more than a strictly logistical and technical meaning, it must describe a decentralized community that allows for direct popular administration, the efficient return of wastes to the countryside, the maximum use of local resources—and yet it must be large enough to foster cultural diversity and psychological uniqueness. The community, like its technology, is itself the ensemble of its libertarian institutions, humanly-scaled structures, the diverse productive tasks that expose the individual to industrial, craft, and horticultural work, in short, the rounded community that the Hellenic polis was meant to be in the eyes of its great democratic statesmen. It is within such a decentralized community, sensitively tailored to its natural ecosystem, that we could hope to develop a new sensibility toward the world of life and a new level of self-consciousness, rational action, and foresight.[20]

The intention of social ecologists in building ecocommunities is to overcome the antimonies of theory and practice at the core of corporate capitalist economies: mind and body, town and country, factory and farm, mental and manual work, consumption and production.

Likewise, this transformation would demand the popular utilization of a new ecological technology,

> composed of flexible, versatile machinery whose productive applications would emphasize durability and quality, not built-in obsolescence, and insensate quantitative output of shoddy goods, and a rapid circulation of expendable commodities. . . . Such an ecotechnology would use the inexhaustible energy capacities of nature—the sun and wind, the tides and waterways, the temperature differentials of the earth and the abundance of hydrogen around us as fuels—to provide the ecocommunity with non-polluting materials or wastes that could be easily recycled.[21]

This ecotechnology is a central goal inasmuch as it implies a less destructive approach toward nature, as well as the more active exercise of individual initiative, self-reliance, and judgment in technological labor. Ecological technology, again, would broaden the choices and freedoms of individual households and communities in their experience of local citizenship and individual producership by reengineering society's total industrial metabolism in Nature.

Social ecology holds that the direct action of individuals could overturn, bit by bit, the administrative regime of centralized welfare states and transnational firms. The abstract machines of global consumption need to be reorganized in concrete machineries of local production by the communities that depend on them. "Ecocommunities and ecotechnologies, scaled to human relationships," Bookchin maintains, "would open a new era in face-to-face relationships and direct democracy, providing the free time that would make it possible in Hellenic fashion for people to manage the affairs of society without the mediation of bureaucracies and professional political functionaries."[22] As a result, Bookchin argues that social ecology would reintroduce and reinforce the moral ends of simpler, more frugal living into the economic struggle against necessity.

In an openly utopian vein, Bookchin concludes that such confederalist political forms could promote a significant transformation:

> The antagonistic division between sexes and age-groups, town and
> country, administration and community, mind and body would be
> reconciled and harmonized in a more humanistic and ecological syn-
> thesis. Out of this transcendence would emerge a new relationship be-
> tween humanity and the natural world in which society itself would
> be conceived as an ecosystem based on unity in diversity, spontaneity,
> and non-hierarchical relationships. Once again we would seek to
> achieve in our own minds the respiritization of the natural world.[23]

By generating a new substantive rationality from the ecological needs of
communities of living things coexisting in the biosphere, a more balanced, equitable, and democratic exercise of instrumental rationality
could emerge from the many diverse, local ecocommunities and their
bioregionally appropriate ecotechnologies.

Exactly how these groups could become a potent political agency in
all spheres of action—urban and rural, mental and manual labor, industrial and agricultural production—remains very vague. Without becoming an ecological vanguard, they ultimately would have to settle on an
almost secessionist notion of "hollowing out" a new society from within
the old one by disengaging individuals, households, and communities.
Once they secede, the political economy of social ecology would have to
scramble in order to redefine, balance, and reintegrate economic needs
with moral necessities. In this ecological "counterstate" seeking a form
of dual power, the obligations of citizenship also would extend into
questions about the use of technology, tools, and the organization of
production. Ultimately, most technical institutions would need to be
treated as accessible communal utilities rather than private properties or
corporate resources.

Like Marcuse, Schumacher, or Soleri, Bookchin believes that the
technology and organizational forms most suitable for an ecommu-
nity are "a technology with a human face, which instead of making
human hands and brains redundant, helps them to become far more
productive than they have ever been before."[24] He notes that this technology is "the basic structural support of a society; it is literally the frame-

work of an economy and of many social institutions."[25] Transnational corporate production plainly has stressed technologies over the past four decades that organize the unecological, antihuman, and amoral social institutions of consumer society. A liberatory ecotechnology, in turn, could provide the new basic structural support for society that is more ecologically sound, humanly scaled, and morally grounded. Following Bookchin, this decentralization of technological power could promote democratization. Under an ecotechnology, not all societies' economic activities

> can be completely decentralized, but the majority can surely be scaled to human communitarian dimensions. This much is certain: we can shift the center of economic power from national to local scale and from centralized bureaucratic forms to local, popular assemblies. This shift would be a revolutionary change of vast proportions, for it would create powerful economic foundations for the sovereignty and autonomy of the local community.[26]

An ecotechnology can be closely integrated into the local environment and the larger biosphere. Bookchin argues that an ecotechnology is morally worthwhile, ecologically necessary, and actually already present in an undeveloped form in corporate science and technology. It exists as a technical possibility, but it is not an inherently necessary change mandated by the unfolding course of history. It must be made to happen by conscious social choices and political action. Ecotechnology, therefore, has become ecologically necessary. Hence, it must not be reduced to purely instrumental technique. Rather, ecotechnology must be seen as this complex social ensemble of techniques, bioregion, and communal productivity. "This ensemble," Bookchin claims, "has the distinct goal of not only meeting human needs in an ecologically sound manner—one which favours diversity within an ecosystem—but of consciously promoting the integrity of the biosphere. The Promethean quest of using technology to 'dominate nature' is replaced by the ecological ethic of using technology to harmonize humanity's relationship with nature."[27]

Like Marcuse, Bookchin attacks the operations of the abstract machines driving the one-dimensionalization of contemporary life, which increasingly integrates all human/machine, human/animal, human/plant, and human/earth interactions into human/human relationships that reduce the organic world into inorganic deadness. He refuses to continue submitting to these abstract machines with their abstracted machinists, who derive disproportionate benefits or dodge exorbitant costs by fabricating our synthetic environments out of nature. Ecotechnologies, like Soleri's designs for arcologies, could help everyone to act as concrete machinists to rethink and reenact technical processes in ecological terms.

The industrial metabolism of global exchange will never become ecologically sustainable until bioregionally sensible communities and Earth-wise individuals politically insist on producing their own economies and cultures outside of the currently installed abstract machineries of global corporate commerce. Liberating human beings from the domination of abstract machines installed as the key servers for the environmentality of huge nation-states and global corporations could, in turn, liberate nonhuman beings from their submission to human-designed and -engineered "economic development" programs. All the problems of modernity will not be solved by Bookchin's solutions, but modernity holds many mazeways for its actual enactment culturally, economically, and politically that opposition to the subpolitics of corporate technologies and national economies could very well revitalize by any attempt to create an ecological order beyond sustainably developed corporate environmentality.

Conclusion: New Departures
for Ecological Resistance

Coming to conclusions is never easy, especially when contesting the politics of nature, economy, and culture. My critical encounters with this diverse collection of contemporary ecocritics may make it even more difficult to draw definitive conclusions about Nature, or politics, ecology, and culture today, inasmuch as my criticisms question comfortable articles of faith advanced in the ecocritiques of these green thinkers. After surveying this selection of ecological social movements, environmental action groups, and green philosophers, the vision of Nature underpinning Biosphere 2 uncomfortably seems to be the most true. Marcuse, Bookchin, and Soleri, as well as Devall, Foreman, and Brown, are all correct: Nature increasingly is no longer a vast realm of unknown, unmanageable, or uncontrollable wild nonhuman activity. After becoming completely ensnared within the megamachinic grids of global production and consumption, as chapter 5 suggests, Nature is turning into "Denature." Much of the earth is a "built environment," a "planned habitat," or "managed range" as pollution modifies atmospheric chemistry, urbanization restructures weather events, architecture encloses whole biomes in sprawling megacities, and biotechnology reengineers the base codes of existing biomass.[1]

In ways suggested by Biosphere 2, the informationalization of industrial and agricultural production is, implicitly and unintentionally, operating as a terraforming exercise, which has reshaped the ecologies of most nation-states to maintain the unsustainable development of transnational commerce. Over the past three or four decades, these changes have become so pervasive and profound that we perhaps must, as Latour argues, amend the key articles in our "discursive constitution," or that set of theoretical and practical principles that "defines humans and non-humans, their properties and their relations, their abilities and their groupings."[2] As first nature constructions of Nature, privileged by deep ecologists and Earth First!, become overlaid with second or third nature constructs, preferred by the Worldwatch Institute or Biosphere 2, the boundaries between natural objects and human subjects blur.[3] High-technology cultures, economies, and societies are producing hybrid agencies and structures, commingling elements of the natural and the artificial, just like Biosphere 2 or Arcosanti, all over the planet. Not unlike Baudrillard's Disneyland, Biosphere 2 now may well be a vital hyper-reality generator, concentrating within its structures the most accurate representations of how Biosphere 1, or the earth, actually works rather than merely serving as the experiment of Biosphere 2.[4] The realities of contemporary environmentality generate new seminatural/semiartificial biomes where quasi objects and quasi subjects amalgamate elements of nature, economy, and culture in ways that deep green ecologists decry and arcological biospherian engineers celebrate.[5]

Ecocritique and Environmentality

In some sense, all of the ecocritiques examined here pose as contra-governmental alternatives, challenging the ways in which the govern-mentality of the current economic and social regime enforces its destructive disposition of things and people in "the environment." To counter such governmentalized interventions in the economy and society, many ecocritics advance their contragovernmental projects to readjust single individuals and/or large human populations to fit more fairly into their visions of change for our environmental setting. Thus, every ecocritique, including the one unfolding here, becomes an expression of another alternative environmentality with its own new codes of ecoknowledge and systems of geopower, articulating fresh theoretical and practical answers to their respective appraisals of the ecological crisis.[6]

Once Nature becomes environmentalized, its workings are submitted to the competing environmentality designs of various ecocritics. Our environmentalized Nature, deformed and re-formed simultaneously in transnational cycles of economic production and cultural reproduction, is the anthropogenic environment—"wilderness preserves" and "sacrifice zones," "national parks" and "industrial belts"—constructed from the practices of sustainable development, green consumerism, global world-watches, or biospheric engineering.[7] Restorational environmentalities, like those of deep ecology or Earth First!, would disintegrate contemporary corporate exchange, returning these processed natural sites back to a future of wild earthiness and human beings to premegatechnical primitivism. Preservational environmentalities, such as The Nature Conservancy or the Worldwatch Institute, would adjust people more rationally to things in their vision of sustainable use for preserved Nature, while transnational commerce continues its projects of sustainable development by slowly consuming such nature preserves. Conservational environmentalities, such as green consumerism or sustainable developmentalism, tout the importance of more efficiently disposing of things used by people and policing global production with strategic environmental initiatives to root out inefficiencies in society's industrial metabolism. And, finally, transformational environmentalities, such as those envisioned by Marcuse, Soleri, or Bookchin, would reconstitute the environment consciously as an emancipatory anthropogenic formation by intervening with new aesthetic sensibilities, arcological alternatives, or ecotechnological devices that might refashion denatured Nature beautifully or spiritually to suit more "aesthetic" visions of ecology. The common conceit of all ecological politics is this aspiration of preserving Nature. Most "green" theories and practices, however, increasingly center on "grey" outcomes—who will "denature" Nature for whom, in what ways, for how long, to serve what ends?—although they often raise these issues very ineffectively with few original insights into what is really unfolding here. Hence, ecocritics propound their various discursive projects with many alternative forms of environmentality to guide individual and collective action in such manufactured ecologies.

At the end of the day, a definitive grasp on the nature of Nature mostly is unattainable for human beings. In the open breach left behind, both science and philosophy construct conventionalized readings of Nature, as these ecocritics illustrate, within codes of ecoknowledge that vary in the emphasis they place on first nature, second nature, or third nature constructions of Nature's essence and appearance.[8] Some ecocritics, such as the deep ecologists, Earth First!, or The Nature Conservancy, stress a first nature in which ordinary terrestriality becomes textualized for deep green interpretations, fundamentalist first readings, or caring conservative constructions of an ecocentric imperative. Other ecocritics, such as the Worldwatch Institute, sustainable developmentalists, or Biosphere 2 designers, emphasize a second nature in which territorialized governmentalities wright terrestrial zones as they write the superscripts of economies, societies, and cultures over the original terrestrial text. From these elisions of secondary superscripts with primary subscripts, environments still form out of Nature/Denature around the anthropole of human communities, justifying anthropocentric guardianship over terrestrial processes to ensure the sustainability of territorialized human communities and their terrestrialized nonhuman ecological services.

Capitalist economies in societies with high technologies, as Bookchin and Marcuse observe, restructure Nature because they concretize abstract technological potentialities in concrete economic structures that make meanings, concentrate energies, form matter, or transform information as "environments" that either suit urbanizing productivism or suburban consumerism. Social ecology essentially is an invitation to contest, evaluate, and transform the subpolitical tendencies, quasicultural distortions or paraeconomic forces in technics that prevent "new sensibilities" of nature, economy, and culture from emerging. Marcuse believes that they are lying dormant within modern science and technology, but powerful blocs of owning interests and ruling agents restructure Nature as systems of domination in their own historicized deployments of technology. Social ecology looks for subpolitics in a politicized technics, quasi cultures in deculturalized institutions, and paraeconomies in noneconomic practices that advance the domination of people by the ways Nature is dominated instrumentally as Denature. Armed with this knowledge, as Soleri suggests, local communities might create their own emancipatory environmentalities—beyond the ruling agendas of suburban consumerism now using scientific disciplines and technical formations to discipline society scientifically and form economies technically—by reinventing the interplay of technics and ecology in their unbuilt and built environments.

All of the ecocritics considered in this book are, in their own ways and on their own terms, working on these problems. Deep ecologists essentially hope to revitalize Nature by disengaging its spaces and processes from the abstract machines of transnational corporate commerce by simply "letting it be." Earth First! pushes this revivalistic agenda further by advocating a serious reconsideration, and even widespread acceptance, of the economies and societies made possible by interacting with Nature on Stone Age technical terms. The personal computer gives you IBM and Microsoft, Earth First! and deep ecology give you the monkey wrench and the stone ax. The Nature Conservancy sees the abstract demands of commodification as relentless and omnipotent, but it need not be omnipresent. Hence, it works to rezone regions of Nature for nonuse now as preparation for future uses, anticipating subsequent but more sophisticated iterations of technology's abstract machineries.

Likewise, the Worldwatch Institute essentially poses as an alternative technology for steering and monitoring the abstract machines of growth, which are now dysfunctional inasmuch as their uncoordinated deployment by self-interested national economies ignores the environmental necessities of transnational planetary economics in the governmental domain. Instead of checking its advances, however, worldwatching environmentalities perhaps promise only to rerationalize their workings by dividing Nature's resources from Denature's assets on the earth. Natural resources would become fixed capital worthy of long-run preservation, while Denature assets would be working capital suitable only for short-run exploitation. Biosphere 2 aspires to restructure the earth's denatured environment as this Denature's earth science technology, turning ecology into a system of terraforming technics. The subpolitics, quasi cultures, and paraeconomics of existing technologies are accepted almost at face value to reconstruct a totally new artificial nature modeled fully on the domination of humans and nonhumans as it is formed by the machinic networks that interlace them together. Soleri, on the other hand, sees the contradictions in the terraforming logic of today's urban form, and pits his arcologies against the subpolitics of consumerism, the quasi cultures of automobility, and the paraeconomics of land abuse embedded in ordinary urban planning, conventional architecture, and traditional building techniques.

Exploring Alternatives: Bookchin, Soleri, Marcuse

Beyond the spectacular democracy, corporate urbanism, and perpetual growth economies of contemporary service states that Marcuse, Bookchin, and Soleri see underpinning the current social consensus, are there thicker or more maximal attributes of community that might be reconstituted in more ecological and populist forms? These questions, and answers to them, call for some imaginative speculation, as Soleri's troubled Arcosanti experiments indicate, if we are even to begin constructing a useful discourse about them. As active citizens in local political associations and as autonomous proprietors in flexible regional markets, might not people work in common at particular places linked by common ecological, economic, and ethical interests? That is, as Marcuse recognizes, by reconstructing the built environments of the city and countryside so that workplace and residence sites, production and consumption processes, administration and allocation interests no longer embody disciplinary decisions made by sustainable developers, resource managerialists, and technical experts but rather those of local communities, might not new communal associations develop that would be open to equal participation by every community member in shared action to make collective decisions, produce a commonwealth, or enjoy mutual benefits? It will require a great deal of imagination, but realizing some of these changes, as Soleri indicates, would not be impossible.

Ecological populist communities could be united by new narratives of their own historical consciousness, beyond and behind the nationalist myths of new class bureaucrats or the progress discourses of corporate public relations, and by more concretely articulated social, ecological, and cultural interests grounded in immediate environmental and political conditions. In continually affirming the project of "history" as such, Bookchin recognizes the significance of this intellectual force for constituting a humane civic order. This means, in turn, that each municipal order must invent new definitive historical stories about its politics, culture, and economy within a particular bioregion to fit its local aspirations and constraints. Whether we characterize it as ecological populism, postscarcity anarchism, arcological living, social ecology, outsider revolution, or democratic municipalism, such changes imply, just as

Soleri and Marcuse suggest, rebuilding the contemporary city and society from the ground up around new aesthetic sensibilities. As common groups sharing particular beliefs, living together in ecologically appropriate built environments, holding equitable shares of property, power, and privilege in the community commonwealth, and practicing more responsible, frugal, and autonomous codes of personal conscience, these thicker communities could move far beyond the passive consumption and dependent clientage now regarded by the new class as a "high standard of living."[9] The crises of everyday material life over the past twenty to twenty-five years continue to unmask this myth of economic progress: it is not standard, it never has been very high for most people, and it may not really be living.

Therefore, each maximal community might find its own ecologically appropriate ways of localistic living, redefine its own self-produced standards of life, and set its own morally responsible ranges of high and low in providing a sustainable living to each and every one of its members. The prevailing new class codes of state-corporate systems of power and knowledge ask communities and individuals to "think globally, act locally" to prop up the tottering political economy of advanced industrial society, but a workable ecological populism ironically will need communities and individuals that can truly "think locally" in new historical narratives and absolutely "act globally" within their own immediate locality.

Most important for an ecological populism, Nature in these communities must be brought back directly into the locality's everyday life from its current denaturalized conditions in more subjective aesthetic and ethical forms. Rather than being only an alien denatured entity from which society rips resources and into which it dumps wastes, Nature should be treated as an equal vital presence in the course of the community's survival.[10] It should not remain the object of administration by new class experts, but the most basic subjective site defining many of the community's essential ends and basic values, such as responsibility, frugality, autonomy, sustainability, or freedom. By being part of the community, its needs for a certain level of care and respect, if only to assure its continuing sustainability, could be more closely safeguarded.

These ecological and populist notions of community must not be mistaken for utopian socialism, rusticated romanticism, or impractical communalism, seeking to live in organic oneness with Nature. Such charges, as Marcuse and Soleri have discovered, will be leveled at any project that aims to push beyond existing forms of economic production, political control, and social organization. Returning to some idyllic natural past, however, is neither likely nor necessary—a point that Bookchin constantly reiterates in his attacks on the uncritical biocentrism taking root in today's progressive ecological organizations. An ecological populism would mean developing an alternative maze within modernity by making different choices about community, or forcing new popular relations of production on the material application of today's misused productive forces, rather than a return with Earth First! to the Pleistocene era for some premodern perfection.[11] Critics of ecological responsibility and populist politics will make bogus claims here about romanticism or conservatism to discredit popular efforts to choose and act differently, but they misinterpret the nature of both the choices and their possible results.

Still, ecological populists must ask different questions to arrive at an ecological economy, society, and polity beyond the collectivism or romanticism of existing ecosocialist rhetoric. All too often, ecosocialist programs seem like agendas of reempowerment for some kind of centralized statist collectivism, which has looked to Maoist peasant work brigades or Cuban state farms for models of "a kinder, gentler socialism" serving green goals.[12] Yet, the collective mode of landholding, work organization, and technological control of such models threatens a disappointing rerun of some bureaucratic *nomenklatura* behind a green facade for some apparat authoritarianism with little further development of individual subjectivity in more emancipatory forms. Similarly, the exciting allure of alternative economies and appropriate technologies in ecosocialism often evokes new ill-considered infatuations. When viewed through the soft focus of *Mother Earth News* or the warm glow of the *Whole Planet Earth Catalogue*, ecosocialism bizarrely parodies Lenin's declaration that "socialism equals electrification plus soviet power" with claims that "ecosocialism equals windmills, geodesic domes, and solar panels plus nonhierarchical, politically correct decision making." This is neither what Bookchin had in mind for his designs for developing an ecotechnology nor what Soleri sees as the essence of arcology. The technological enthusiasms of such "back to the land, here comes the sun" manifestos unfortunately also promise to exhaust their force in economic fragmentation, cultural frustration, and political failure. An ecological populism informed by Marcuse's new aesthetic sensibilities should be practical, workable, and realistic, pushing its agendas far beyond this kind of failed ecosocialist millenarianism.

Ecological Populism and Communalistic Localism

An ecologically oriented popular rule guided by a new aesthetic sensibility and sited in a community-built arcology has been touched on tentatively by Bookchin, Marcuse, and Soleri. For all of them, the emancipatory realization of real face-to-face participatory democracy and sustainable autonomous proprietorship within vital local communities must realize greater personal and communal freedom so that ecological transformations could represent, in practice, a radical political advance rather than a regressive social reaction. Most important, these three theorists also believe that economies based on feudal manorialism or neolithic hunting and gathering are not viable possibilities for contemporary social formations within either the existing global economy or the prevailing man-made environment, despite the arguments made on their behalf by biocentrist deep ecology. They also would dispute the notion that a "future primitivism" would be the inevitable result of successful popular movements today against new class domination; indeed, these scare tactics only show a complete lack of imagination on the behalf of its conservative supporters.

By remaining attuned to Bookchin's ecological imperatives of sustainability, populist communities in a Solerian arcology with their new ecotechnologies would not need to buy into the prevailing liberal codes of possessive individualism. Similarly, in accord with Marcuse's new sensibility, their expectation of greater individual autonomy from each person and every household could disconnect many of these communities' consumeristic links into major corporate modes of consumption in favor of building indigenous economies of ecologically responsible production and consumption. "Thinking locally" and "acting globally" would mean making several radical shifts into styles of operation, modes of organization, and ways of action conventionally regarded as "dead and gone." Yet, as Bookchin's classical notions of ethical action and Marcuse's brief on historical self-understanding maintain, the conventions of anticommunitarian technocracy ordinarily used to fill out this cultural death certificate or final shipping manifest have been manufactured by new class experts as yet another piece of the puzzle behind their power and privilege.[13] Thus, one need not be troubled too deeply by the self-interested reactions of new class critics, who are threatened by the loss of authority and responsibility that would follow from any popular ecological movements pushing for a more direct empowerment. Defining the more concrete policy changes for realizing this sort of ecological revolution is a much more speculative pursuit, but it still is worth beginning.

First, "acting globally" in this context may mean creating more complex, diverse, and skilled societies of small producers, who own a considerable amount of real property and control a respectable body of valuable skills. Owning and controlling such assets directly at the local and regional level would be the first step toward reunifying production and consumption in the same population centers. This means changing some significant conditions in the abstract processes of social mechanization, such as who owns and controls as well as who uses and profits from land, capital, and technology. These are major changes, but essential ones. By making those who consume aware of the ecological, economic, and energy costs of modern production, as well as those who produce aware of the ecological, economic, and energy costs of contemporary consumption, the modern system of creating and using wealth might change by finding a new modernity, not returning to premodernity. Not only the benefits, but also the costs, would then be evident everywhere at the local level rather than only in distant "sacrifice zones," "pollution havens," or "agribusiness regions." A complex society of small-scale proprietors and regional producers, in turn, could nest their economies more ecologically in the particular ecoregion that they inhabit and must keep sustainable.[14]

Second, "acting globally" here might imply cultivating a new subjectivity grounded on new kinds of empowerment—technological, economic, political, and cultural. Since "the good life" would no longer be the endless consumption games of contemporary permissive individualism, it must be redefined in more demanding moral codes of hard work, frugality, ecological responsibility, humility, and skill perfection.[15] As owners and traders working within a particular locality or region, every individual's technical knowledge and economic labor would have to sustain his or her family and other families. Without his or her skilled work, the community would suffer, but with his or her frugal responsibility each community and ecoregion should prosper sustainably. Although everyone would enjoy a more modest material "standard of living," at the same time, wealth, skill, and property should be distributed much more equitably to give everyone both a stake and a share in the sustainable economies of small-scale production. Many abuses now occur because few people have a meaningful share, and even fewer possess a significant stake, in the new class's global economic system. Making the effects of environmental destruction and preservation more personal and communal could lessen many ecological disasters happening right now.

Third, "acting globally" may require new community confederalist institutions, which are constructed by, for, and of the particular peoples of each locality to suit their smaller scale of communal production, governance, and development.[16] The truly progressive advances made by secular rational civilization might be counterbalanced against the potentially regressive return to reactionary irrational culture. Small-scale communal production made fairly universal on a national or global scale will, nonetheless, remain tied together to outside communities through existing telecommunication and transportation systems. Such technologies are unlikely to be disinvented even as their use is made more popular. Racism, provincialism, xenophobia, sexism, and class hatreds need not necessarily be part of any particular ecological society, or, at least, no more a part of them than they are now within the contemporary nation-state. Loyalties to community, ecoregion, or place could become, but should not become, lines of cultural conflict or group warfare, especially if Marcuse's new sensibility were taken seriously. The systems of communal production, as well as the structures of community democracy, ought to reinforce a culture of fairness, toleration, and humility as they develop together with other communities seeking similar ends.

Fourth, "acting globally" might necessitate lessening the scope of centralized bureaucratic state control and curbing the sweep of standardized corporate penetration in local communities, while increasing the levels of responsibility and activity exerted by each individual living in these decolonized local communities. The "downsizing" of nation-states and "hollowing out" of major firms already anticipates this condition, but it could go much further. As Bookchin and Soleri indicate, self-rule, self-ownership, and self-management would be rigorously demanding practices. And, freedom will entail the prospect of failures, reversals, and just having less at times without having the security of turning automatically to centralized state powers for relief. Similarly, within ranges set out by some generally agreed upon guidelines, communities also must be allowed to hold and practice more diverse sets of local cultural values once they are left to their own local and communal devices. Just as they may largely succeed at making a popular ecological community work, they also might completely collapse. This largely untrammeled freedom to fail makes their autonomy more meaningful. If some bloc of bureaucratic big brothers is always standing in the background ready to pick them up, dust them off, and push them along their way, then there will be very little meaningful autonomy for such populist commonwealths.[17] The transnational qualities of the global economy with its time-and-space compressing communication and transportation systems will prevent most communities from withdrawing into isolationist self-sufficiency or completely collapsing after some unforeseen natural disaster. Still, each community, in order to be free to realize success, must also be left open to suffer from failure.

Finally, "acting globally" may involve reconstituting the fundamental writs of authority now underpinning public order. Propounding larger aggregates of these autonomous communities in the United States, or any comparatively well-functioning nation-state, will require new kinds of confederal structures to protect and preserve such entities from outside interference, foreign conquest, and internal insecurity. An entirely new discourse of democratic constitutional founding, which can reconstitute collective images of community, culture, and commonwealth around ecological permanence, to suit the political goals of these communities, will be required. They must be disentangled, first, from the somewhat bankrupt nationalist myths of ethnocentric founding for sovereign authority used by most centralizers, unionists, or bureaucratizers, including many of today's resource managerialists, to defend "the perpetual union" of these United States and, second, from the always suspect state myths of founding and sovereign authority used by most decentralizers, secessionists, or popularizers to extol "states' rights" within the United States. Where nation-states are not yet operating successfully, the prospects for such bioregional or municipal regimes might make good governance a more real possibility.

Such communitarian transformations will be very difficult. They represent a complete revolution in the abstract machines used now to organize everyday life. Talking about community control, state's rights, or local choice in the United States, for example, immediately raises issues of race and racial exploitation, which have, as the environmental justice movement shows, real ecological implications. Nonetheless, fresh discourses about new community-centered alternatives must center on local autonomy without revitalizing traditional practices of racism, sexism, or classism. Preserving national and transnational ties for these communities also must not aggravate liberal practices of unrepresented rationalization, arbitrary bureaucratic decision making, and collective organization without popular consent.[18] By moving forward to test such unrealized future possibilities, an ecological populism could shake many communities' traditional reactionary resistance to any kind of change, as well as organize political countermeasures against new class domination from large nation-states and major corporations in everyday life.

Ecocritique as Visions of Alternative Modernities

By optimizing large-scale economies of energy intensity, planned efficiency, capital intensivity, and global marketing, as many of the ecocritics examined here have argued, the new class from its offices with major nation-states and transnational capital has rebuilt the world economy since 1945 along ecologically destructive lines. A localistic ecological populism, as a transformative social project, seeks the technological and organizational means to rebuild this global corporate order along much different institutional lines: small-scale, energy-sensible, locally managed, labor intensive, bioregionally structured communities of economic autonomy. Accepting equal partnership with Nature instead of the domination of denaturalized environments frames the moral purposes behind living within the terms of an ecotechnology. Living more simply with the biosphere's cycles of generation and regeneration might make it possible to design social arcologies as a new moral-political order to preserve the ecocommunities they would nurture. The reduction of Nature's domination also should reduce, as Marcuse and Bookchin argue, human domination of other humans. Protecting Nature would make all human minds and hands useful in a new system of "production for the masses" based more fully on satisfying nonmaterial ends. As Schumacher suggests in another context, a bioregional or local ecological populism should mobilize

> resources which are possessed by all human beings, their clever brains and skillful hands, *and supports them with first-class tools.* The technology of mass production is inherently violent, ecologically damaging, self-defeating in terms of non-renewable resources, and stultifying for the human person. The technology of *production by the masses*, making use of the best of modern knowledge and experience, is conducive to decentralisation, compatible with the laws of ecology, gentle in its use of scarce resources, and designed to serve the human person instead of making him the servant of machines.[19]

By using these material means, Bookchin, Soleri, and Marcuse point out that ecocommunities would more easily realize in everyday life those vital nonmaterial ends—justice, balance, frugality, enjoyment—that most persons would seek in an ecological community.

Such tools, once positioned in the hands of ecological populists, might also enable confederalized ecological arcologies to practice more "convivial" values—self-expression, individual choice, household autonomy, community empowerment, bioregional subsistence.[20] In a marked contrast to today's corporate marketplace, Bookchin's social ecology in confederations of Soleri's arcologies could revitalize many almost lost values, such as "thrift, simplicity, diversity, neighborliness, humility, and craftsmanship,"[21] implicit in Marcuse's new sensibility, in caring for the biosphere. Most modern tools contain unrealized, but still quite real, self-subversive dimensions that ecotechnologies wielded by aesthetically minded citizens could tap to meet the new arcological purposes of eco-communal development. The inaccessible sciences of new class corporate and state specialists could be reconstituted as popular traditions of accessible communal knowledge useful to nonspecialists.[22] From a technical vantage, it is quite possible to design instruments that are simple, durable, useful, and manageable in either individual, household, or community use as alternative sets of abstract machines capable of re-modulating human/machine, human/animal, human/plant interactivities in unprecedented ecological configurations. A new sensibility, a new technology regime, and a new process aesthetics all could develop then along with this sort of ecological community.

To conclude, modernity is not a unilinear and irreversible course from primitive community to complex society, despite what mainstream sociological theory or new class bureaucrats would have everyone believe.[23] As Bookchin argues, it is both the result and the practice of continuously making critical choices under certain economic, ideological, and organizational constraints set by the use of power and knowledge in a context where elements of the past, present, and future mutually co-exist. It does not need to be the way that it has come to be. New technologies created by new communities can generate new individual and collective subjectivities within new environments that are far less like the wasteland of today's Denature and much more like some other day's Nature. There are many alternative modernities that can be created by making different, more popular, choices about how production might be organized, where communities are centered, when power is exercised, who could participate, and why subjectivity should change. The first steps toward getting there, no matter how contradictory or confused, are being taken by these ecocritics in their ecocritiques.

Notes

Introduction. Contesting the Politics of Nature, Economy, and Culture

1. The greatest and first of these writings, of course, is Rachel Carson, *Silent Spring* (Boston: Houghton Mifflin, 1962), although Murray Bookchin, writing under the pseudonym of Lewis Herber, and his *Our Synthetic Environment* (New York: Knopf, 1962), must be mentioned as another significant influence in the early 1960s. Histories of ecology as a discipline can be found in Donald Worster, *Nature's Economy: The Roots of Ecology* (Garden City, N.Y.: Anchor Books, 1979); Robert P. McIntosh, *The Background of Ecology: Concept and Theory* (New York: Cambridge University Press, 1985); or Anna Bramwell, *Ecology in the Twentieth Century* (New Haven: Yale University Press, 1989). During the past three decades, the currents of ecological criticism following Carson's lead have carried many different kinds of analysis, including Wendell Berry, *The Unsettling of America: Culture and Agriculture* (New York: Avon, 1977); Murray Bookchin, *Post-Scarcity Anarchism* (Berkeley: Ramparts Press, 1971); Lester R. Brown, *World without Borders* (New York: Random House, 1972); Fritjof Capra and Charlene Spretnak, *Green Politics* (New York: E. P. Dutton, 1984); William R. Catton, *Overshoot: An Ecological Basis of Revolutionary Change* (Urbana: University of Illinois Press, 1980); Barry Commoner, *The Closing Circle: Nature, Man, and Technology* (New York: Knopf, 1971); Bill Devall and George Sessions, *Deep Ecology* (Salt Lake City: Peregrine Smith Books, 1985); Paul Ehrlich, *The Population Bomb* (San Francisco: Sierra Club Books, 1968); André Gorz, *Ecology as Politics* (Montreal: Black Rose Books, 1980); Robert Heilbroner, *An Inquiry into the Human Prospect* (New York: Norton, 1980); William Leiss, *The Limits to Satisfaction* (Toronto: University of Toronto Press, 1976); James E. Locklock, *Gaia: New Look at Life on Earth* (New York: Oxford University Press, 1979); Lewis Mumford, *The Myth of the Machine*, vol. 1, *Tech-*

nics and Human Development, vol. 2, *The Pentagon of Power* (New York: Harcourt Brace Jovanovich, 1967, 1970); Kirkpatrick Sale, *Dwellers in the Land: The Bioregional Vision* (San Francisco: Sierra Club Books, 1985); E. F. Schumacher, *Small Is Beautiful: Economics as if People Mattered* (New York: Harper and Row, 1973); Alvin Toffler, *The Eco-Spasm Report* (New York: Bantam Books, 1975); and Barbara Ward and René Dubos, *Only One Earth* (New York: Norton, 1972). In addition to these ecocritiques, one can find literary scholars and theorists appropriating "ecocriticism" as their own exclusive franchise. In this vein, "ecocriticism is the study of the relationship between literature and the physical environment. . . . ecocriticism takes an earth-centered approach to literary studies" (Cheryll Glotfelty, "Introduction: Literary Studies in an Age of Environmental Crisis," *The Ecocriticism Reader,* ed. Cheryll Glotfelty and Harold Fromm [Athens: University of Georgia Press, 1996], xiii).

2. For some conventional reevaluations of these tendencies, see Philip Shabecoff, *A New Name for Peace: International Environmentalism, Sustainable Development, and Democracy* (Hanover, N.H.: University of New England Press, 1996); Mark Dowie, *Losing Ground: American Environmentalism at the Close of the Twentieth Century* (Cambridge, Mass.: MIT Press, 1995); John O'Neill, *Ecology, Policy and Politics: Human Well-Being and the Natural World* (London: Routledge, 1993); Philip Shabecoff, *A Fierce Green Fire: The American Environmental Movement* (New York: Hill and Wang, 1993); Peter Dickens, *Society and Nature: Towards a Green Social Theory* (Philadelphia: Temple University Press, 1992); Robyn Eckersley, *Environmentalism and Political Theory* (Albany: State University of New York Press, 1992); Andrew Dobson, *Green Political Thought: An Introduction* (London: Unwin Hyman, 1990); John Young, *Sustaining the Earth* (Cambridge, Mass.: Harvard University Press, 1990); and Samuel P. Hays, *Beauty, Health, and Permanence: Environmental Politics in the United States, 1955–1985* (Cambridge: Cambridge University Press, 1987).

3. Some significant recent examples would include Tom Athanasiou, *Divided Planet: The Ecology of Rich and Poor* (Boston: Little, Brown, 1996); Ulrich Beck, *Ecological Politics in an Age of Risk* (Cambridge: Polity Press, 1995); Thomas Berry, *The Dream of the Earth* (San Francisco: Sierra Club Books, 1988); Bill McKibben, *The End of Nature* (New York: Random House, 1989); Barry Commoner, *Making Peace with the Planet* (New York: Pantheon, 1990); Jeremy Rifkin, *Biosphere Politics: A New Consciousness for a New Century* (New York: Crown, 1991); Al Gore, *Earth in the Balance: Ecology and the Human Spirit* (Boston: Houghton Mifflin, 1992); Garrett Hardin, *Living within Limits: Ecology, Economics and Population Taboos* (New York: Oxford University Press,

1993); and Max Oelschlaeger, *Caring for Creation: An Ecumenical Approach to the Environmental Crisis* (New Haven: Yale University Press, 1994).

4. For a sample of some recent attempts to systemically discuss the interplay of environmental concerns and contemporary politics in the United States, see Koula Mellos, *Perspectives on Ecology: A Critical Essay* (New York: St. Martin's Press, 1988); Robert Paehlke, *Environmentalism and the Future of Progressive Politics* (New Haven: Yale University Press, 1989); Bob Pepperman Taylor, *Our Limits Transgressed: Environmental Political Thought in America* (Lawrence: University Press of Kansas, 1992); Robert Gottlieb, *Forcing the Spring: The Transformation of the American Environmental Movement* (Washington, D.C.: Island Press, 1993); and Kirkpatrick Sale, *The Green Revolution: The American Environmental Movement 1962–1992* (New York: Hill and Wang, 1993). For a more positive spin on the ecological crisis, see Gregg Easterbrook, *A Moment on Earth: The Coming Age of Environmental Optimism* (New York: Viking, 1995).

5. The project of fusing ecology with socialist, feminist, or progressive politics is discussed by Rudolf Bahro, *Building the Green Movement* (London: New Society Publishers, 1986); Daniel A. Coleman, *Ecopolitics: Building a Green Society* (New Brunswick, N.J.: Rutgers University Press, 1994); Penny Kemp and Derek Wall, *A Green Manifesto for the 1990s* (London: Penguin, 1990); Carolyn Merchant, *Earthcare: Women and the Environment* (New York: Routledge, 1995); Mary Mellor, *Breaking the Boundaries: Towards a Feminist Green Socialism* (London: Virago, 1992); Jonathon Porritt, *Seeing Green: The Politics of Ecology Explained* (Oxford: Basil Blackwell, 1985); Jeremy Rifkin and Carol Grunewald Rifkin, *Voting Green* (New York: Doubleday, 1992); Martin Ryle, *Ecology and Socialism* (London: Radius, 1988); Brian Tokar, *The Green Alternative: Creating an Ecological Future* (San Pedro, Calif.: R. and E. Miles, 1987); and Derek Wall, *Getting There: Steps to a Green Society* (London: Merlin, 1990).

6. One useful starting point for rethinking human ecologies in terms of biopower is Michel Foucault, *The History of Sexuality*, vol. 1, *An Introduction* (New York: Vintage, 1980), 81–114. For further discussion of these power/knowledge relations as machinic systems, see Gilles Deleuze and Félix Guattari, *A Thousand Plateaus: Capitalism and Schizophrenia* (Minneapolis: University of Minnesota Press, 1987), 140–43.

7. Some sustained, but often also occasionally overdrawn, criticisms of what many authors now regard as "radical ecology" can be found in Martin W. Lewis, *Green Delusions: An Environmentalist Critique of Radical Environmentalism* (Durham, N.C.: Duke University Press, 1992); Charles T. Rubin, *The Green Crusade: Rethinking the Roots of Environmentalism* (New York: Free Press, 1994); and Michael E. Zimmermann, *Contesting Earth's Future: Radical Ecology*

and Postmodernity (Berkeley: University of California Press, 1994). More insights also can be gained from reviewing the wilderness philosophies that underpin what now is labeled as "radical ecology." For some of these wilderness philosophies, see Roderick Nash, *Wilderness and the American Mind*, 3d ed. (New Haven: Yale University Press, 1982); Arne Nash, *Ecology, Community, and Lifestyle: Outline of an Ecosophy*, trans. and ed. David Rothenberg (New York: Cambridge University Press, 1989); Max Oelschlaeger, *The Idea of Wilderness: From Prehistory to the Age of Ecology* (New Haven: Yale University Press, 1991); and David Rothenberg, ed., *Wild Ideas* (Minneapolis: University of Minnesota Press, 1995).

8. See Edward Abbey, *The Monkey Wrench Gang* (New York: Avon, 1975), and its sequel, *Hayduke Lives!* (Boston: Little, Brown, 1990). A consideration of Abbey's politics and writings can be found in James Bishop Jr., *Epitaph for a Desert Anarchist: The Life and Legacy of Edward Abbey* (New York: Atheneum, 1994).

9. More elaboration of these commitments is developed by Christopher Manes, *Green Rage: Radical Environmentalism and the Unmaking of Civilization* (Boston: Little, Brown, 1990); and Dave Foreman, *Confessions of an Eco-Warrior* (New York: Harmony Books, 1991).

10. For a positive affirmation of these tendencies, see Herman E. Daly and John B. Cobb Jr., *For the Common Good: Redirecting the Economy Toward Community, the Environment, and a Sustainable Future* (Boston: Beacon Press, 1989); Christopher D. Stone, *The Gnat Is Older Than Man: Environment and Human Agenda* (Princeton, N.J.: Princeton University Press, 1993); and Donella Meadows et al., *Beyond the Limits: Confronting Global Collapse, Envisioning a Sustainable Future* (Post Mills, Vt.: Chelsea Green, 1992).

11. See Timothy W. Luke, *Social Theory and Modernity: Critique, Dissent and Revolution* (Newbury Park, Calif.: Sage, 1990), 128–58.

12. Herbert Marcuse, *One Dimensional Man: Studies in the Ideology of Advanced Industrial Society* (Boston: Beacon Press, 1966).

13. Luke, *Social Theory and Modernity*, 159–82. For another discussion linking aesthetic sensibilities and environmental thinking, see Arnold Berleant, *The Aesthetics of Environment* (Philadelphia: Temple University Press, 1992); and Gary Snyder, *A Place in Space: Ethics, Aesthetics and Watersheds* (Washington, D.C.: Counterpoint Press, 1995).

14. Most of the ethical strengths and epistemological weaknesses of Bookchin's projects can be seen at play in his *Ecology of Freedom* (Palo Alto, Calif.: Cheshire Books, 1982).

15. See Tim Luke, "Informationalism and Ecology," *Telos* 56 (summer

1983): 56–73, as well as Tim Luke, "Radical Ecology and the Crisis of Political Economy," *Telos* 46 (winter 1980–81): 63–72.

16. Further discussion of the basic economic and political thinking behind the "wise use" movement can be found in James G. Watt, *The Courage of a Conservative* (New York: Simon and Schuster, 1985). More detailed investigations are developed by Brian Tokar, "The 'Wise-Use' Backlash: Responding to Militant Anti-Environmentalism," *Ecologist* 25 (July–August 1995): 150–56; David Lapp, "Wise Use's Labor Ruse," *Environmental Action* 25 (fall 1993): 23–27; Margaret L. Knox, "The Word According to Cushman: Wise Use Movement Leader Chuck Cushman," *Wilderness* 56 (spring 1993): 28–33; William Poole, "Neither Wise nor Well: Wise Use Movement," *Sierra* 77 (November–December 1992): 58–68; Patricia Brynes, "The Counterfeit Crusade: The Wise-Use Movement," *Wilderness* 55 (summer 1992): 29–32; and Margaret Knox, "Meet the Anti-Greens: The 'Wise-Use' Movement Fronts for Industry," *Progressive* 55 (October 1991): 21–24.

17. For another attempt to criticize these globally articulated megamachines in hope of finding a more ecological alternative for organizing human/ machine, human/animal, human/plant interactions, see Mumford, *The Myth of the Machine*, vol. 2, *The Pentagon of Power*, as well as David Rothenberg, *Hand's End: Technology and the Limits of Nature* (Berkeley: University of California Press, 1993).

18. Worster, *Nature's Economy*, 347–49.

19. As John Elder, professor of English and Environmental Studies at Middlebury College, suggests, literary scholars still are arguing over the conceptual terms of their debate with regard to the environment. Whether it is "green cultural studies," "Nature writing," or "ecocriticism," Elder claims that the fact that "we can't agree on a name is a sign of the incredible diversity in ecocriticism. It's a free-for-all, and that's exciting" (cited in Karen J. Winkler, "Inventing a New Field: The Study of Literature about the Environment," *Chronicle of Higher Education* 42, no. 48 [August 9, 1996]: A8). In the final analysis, my approach to ecocritique, as it is developed in this book, might be seen as a blend of "green cultural studies" and "green political theory." Consequently, my ecocritiques second the assertions of Lawrence Buell, professor of English at Harvard University, who claims "that the worst thing that could happen would be for ecocriticism to become just another branch of literary criticism" (see Winkler, "Inventing a New Field," A15).

20. For some provisional advances in this direction, see Timothy W. Luke, *Departures from Marx: Constructing an Ecological Critique of the Informational Revolution* (Urbana: University of Illinois Press, forthcoming); Kirkpatrick Sale,

Human Scale (New York: Coward, McCann and Geoghegan, 1980); and David Dickson, *Alternative Technology and the Politics of Technical Change* (Glasgow: Fontana, 1974).

1. Deep Ecology as Political Philosophy

1. Some key texts are Arne Naess, *Ecology, Community, and Lifestyle: Outline of an Ecosophy*, trans. and ed. David Rothenberg (Cambridge: Cambridge University Press, 1989); Bill Devall and George Sessions, *Deep Ecology* (Salt Lake City: Peregrine Smith Books, 1985); Bill Devall, *Simple in Means, Rich in Ends* (Salt Lake City: Peregrine Smith Books, 1988); and George Sessions, *Deep Ecology for the 21st Century: Readings on the Philosophy and Practice of the New Environmentalism* (Boston: Shambhala, 1995). See also Alan Drengson and Yuichi Inoue, *The Deep Ecology Movement: An Introductory Anthology* (Berkeley: North Atlantic Books, 1995); Warwick Fox, *Toward a Transpersonal Ecology: Developing New Foundations for Environmentalism* (Boston: Shambhala, 1990); and Christopher Manes, *Green Rage: Radical Environmentalism and the Unmaking of Civilization* (Boston: Little, Brown, 1990). A critique of deep ecology can be found in George Bradford, *How Deep Is Deep Ecology?* (Ojai, Calif.: Times Change Press, 1989); Michael Zimmerman, *Contesting Earth's Future: Radical Ecology and Postmodernity* (Berkeley: University of California Press, 1994); and Luc Ferry, *The New Ecological Order* (Chicago: University of Chicago Press, 1995).

2. Bill Devall, "Ecological Consciousness and Ecological Resisting: Guidelines for Comprehension and Research," *Humboldt Journal of Social Relations* 9, no. 2 (spring 1982): 177–96; Bill Devall, "The Deep Ecology Movement," *Natural Resources Journal* 20 (April 1980): 295–322; and see Dave Foreman, ed., *Ecodefense: A Field Guide to Monkey Wrenching* (Tucson: Earth First! Books, 1985), for a discussion of such radical groups and their tactical styles. See also Rik Scarce, *Eco-Warriors: Understanding the Radical Environmental Movement* (Chicago: Noble Press, 1990).

3. Jonathon Porritt, *Seeing Green: The Politics of Ecology Explained* (Oxford: Basil Blackwell, 1985); and Petra Kelly, *Fighting for Hope* (Boston: South End Press, 1984).

4. See Carl Boggs, *Social Movements and Political Power: Emerging Forms of Radicalism in the West* (Philadelphia: Temple University Press, 1986); and Kirkpatrick Sale, *Human Scale* (New York: Coward, McCann and Geoghegan, 1980). For more discussion of deep ecology's ties to ecological resistance movements, see Bron Raymond Taylor, ed., *Ecological Resistance Movements: The Global Emergence of Radical and Popular Environmentalism* (Albany: State Uni-

versity of New York Press, 1995); and David Pepper, *Modern Environmentalism: An Introduction* (London: Routledge, 1996).

5. Donald Worster, *Nature's Economy* (San Francisco: Sierra Club Books, 1977).

6. Barry Commoner, *The Closing Circle: Man, Nature and Technology* (New York: Bantam Books, 1971), 125.

7. George Sessions, "Shallow and Deep Ecology: A Review of the Philosophical Literature," in *Ecological Consciousness: Essays from the Earthday X Colloquium,* ed. Robert C. Schultz and J. Donald Hughes (Washington, D.C.: University Press of America, 1981), 392.

8. See Stewart Udall, *The Quiet Crisis* (New York: Holt, Rinehart and Winston, 1963).

9. See John G. Mitchell with Constance L. Stallings, *Ecotactics: The Sierra Club Handbook for Environment Activities* (New York: Pocket Books, 1970); Alexander Cockburn and James Ridgeway, eds., *Political Ecology: An Activist's Reader on Energy, Land, Food, Technology, Health and the Economies of Social Change* (New York: Times Books, 1979); and Bill Devall, "Reformist Environmentalism," *Humboldt Journal of Social Relations* 6, no. 2 (summer 1979): 129–55.

10. Devall and Sessions, *Deep Ecology,* 3.

11. Sessions, "Shallow and Deep Ecology," 391–462.

12. Arne Naess, "The Shallow and the Deep, Long–Range Ecology Movements: A Summary," *Inquiry* 16 (spring 1973): 95–100.

13. Ibid., 100; emphasis in the original.

14. Ibid.

15. Devall and Sessions, *Deep Ecology,* 65.

16. Ibid., 65–66.

17. Gary Snyder, *Turtle Island* (New York: New Directions, 1974).

18. Devall and Sessions, *Deep Ecology,* 67.

19. Ibid., 68.

20. Ibid.

21. Cited in ibid., 70.

22. Ibid., 96.

23. Max Horkheimer and Theodor W. Adorno, *Dialectic of Enlightenment* (New York: Seabury, 1972), 3.

24. Ibid., 4–6.

25. Ibid., 6–9.

26. Devall and Sessions, *Deep Ecology,* 66.

27. Michael Tobias, ed., *Deep Ecology* (San Diego: Avant Books, 1985), v.

28. Devall and Sessions, *Deep Ecology*, 97.

29. Arne Naess, "Identification as a Source of Deep Ecological Attitudes," in *Deep Ecology*, ed. Michael Tobias, 256–70.

30. Devall and Sessions, *Deep Ecology*, 67.

31. Ibid.

32. Ibid.

33. Ibid.

34. See Stephen K. White, *Political Theory and Postmodernism* (Cambridge: Cambridge University Press, 1991), 13–30.

35. Devall and Sessions, *Deep Ecology*, 20–21.

36. See William Cronon, *Changes in the Land: Indians, Colonists, and the Ecology of New England* (New York: Hill and Wang, 1983); and Peter Farb, *Man's Rise to Civilization as Shown by the Indians of North America from Primeval Times to the Coming of the Industrial State* (New York: E. P. Dutton, 1968).

37. Devall and Sessions, *Deep Ecology*, 97.

38. See Doris LaChapelle, *Earth Wisdom* (San Diego: Guild of Tudors Press, 1978).

39. Devall and Sessions, *Deep Ecology*, ix.

40. Horkheimer and Adorno, *Dialectic of Enlightenment*, 15.

41. See, for example, James Frazer, *The Golden Bough* (New York: Macmillan, 1953); and Bronislaw Malinowski, *Magic, Science and Religion* (Garden City, N.Y.: Anchor Books, 1955).

42. Horkheimer and Adorno, *Dialectic of Enlightenment*, 8.

43. Ibid., 11–12.

44. See Fritjof Capra, *The Tao of Physics* (Boulder, Colo.: Shambhala Press, 1975); and Morris Berman, *The Reenchantment of the World* (Ithaca, N.Y.: Cornell University Press, 1981).

45. For a positive treatment of New Age thought, see Marilyn Ferguson, *The Aquarian Conspiracy* (New York: St. Martin's Press, 1980). Also see Richard Buckminister Fuller, *Operating Manual for Spaceship Earth* (New York: E. P. Dutton, 1971).

46. The emergence of modern science and its initial coevolution with mystical interests are discussed in Timothy J. Reiss, *The Discourse of Modernism* (Ithaca, N.Y.: Cornell University Press, 1982); and William Leiss, *The Domination of Nature* (Boston: Beacon Press, 1974).

47. See Robert A. Pois, *National Socialism and the Religion of Nature* (New York: St. Martin's Press, 1986).

48. Devall and Sessions, *Deep Ecology*, 100–101.

49. Sessions, "Shallow and Deep Ecology," 412.

50. Max Weber, *The Sociology of Religion* (Boston: Beacon Press, 1964), 267.

51. Devall and Sessions, *Deep Ecology*, 66–67.

52. Ibid.

53. Georg Lukács, *History and Class Consciousness* (Cambridge, Mass.: MIT Press, 1971), 136.

54. Devall and Sessions, *Deep Ecology*, 67.

55. Ibid., 68.

56. Ibid.

57. For more discussion along these lines, see Berman, *The Reenchantment of the World*, as well as Fox, *Toward a Transpersonal Ecology*.

58. Michel Foucault, *Discipline and Punish: The Birth of the Prison* (New York: Vintage, 1979).

59. Aldo Leopold, *Sand County Almanac* (New York: Oxford University Press, 1968).

60. See Marshall Berman, *The Politics of Authenticity: Radical Individualism and the Emergence of Modern Society* (New York: Atheneum, 1970); and Robert Nisbet, *The Quest for Community* (New York: Oxford University Press, 1953).

61. Timothy W. Luke, "On Nature and Society: Rousseau versus the Enlightenment," *History of Political Thought* 5, no. 2 (summer 1984): 211–43.

62. See Jean-Jacques Rousseau, *Émile* (New York: E. P. Dutton, 1911); and Tracy Strong, *Jean-Jacques Rousseau: The Politics of the Ordinary* (Thousand Oaks, Calif.: Sage, 1994).

63. Devall and Sessions, *Deep Ecology*, ix.

64. Karl Marx, *The Marx-Engels Reader*, ed. Robert C. Tucker, 2d ed. (New York: Norton, 1978), 4.

65. Chim Blea, "Animal Rights and Deep Ecology Movements," *Synthesis* 23 (1986): 13–14. For more discussion of Foreman's thinking, see Martha F. Lee, *Earth First! Environmental Apocalypse* (Syracuse: Syracuse University Press, 1995), 62–75.

66. Devall and Sessions, *Deep Ecology*, 188.

67. Edward Abbey, *Desert Solitaire* (New York: Ballantine Books, 1971).

68. See Fred Hirsch, *Social Limits to Growth* (London: Routledge and Kegan Paul, 1977), 21–66.

69. Devall and Sessions, *Deep Ecology*, 70.

70. Chim Blea, "Animal Rights and Deep Ecology Movements," 13.

71. Ibid., 14.

72. See Warwick Fox, "Deep Ecology: A New Philosophy for Our Time," *Ecologist* 14, nos. 5–6 (1984): 194–200.

73. Devall and Sessions, *Deep Ecology*, 14.

74. Arne Naess, "Intuition, Intrinsic Value and Deep Ecology," *Ecologist* 14, nos. 5–6 (1984): 201–4.

75. Devall and Sessions, *Deep Ecology*, 205.

76. Snyder, *Turtle Island*, 99.

77. Ibid.

78. Naess, "Identification," 269–70.

79. Fox, "Deep Ecology," 199.

2. Ecological Politics and Local Struggles: Earth First! as an Environmental Resistance Movement

1. For an extended historical overview of Earth First! see Susan Zakin, *Coyotes and Town Dogs: Earth First! and the Environmental Movement* (New York: Viking, 1993), as well as Rik Scarce, *Eco-Warriors: Understanding the Radical Environmental Movement* (Chicago: Noble Press, 1990). A more engaged discussion of Earth First! ideas is developed in these works: Dave Foreman, *Confessions of an Eco-Warrior* (New York: Harmony Books, 1991); Bill McKibben, *The End of Nature* (New York: Knopf, 1989); and Charles Bowden, *Blue Desert* (Tucson: University of Arizona Press, 1988), 87–98.

2. Charles Bowden, "Dave Foreman!" *Buzzworm* 2, no. 2 (March–April, 1990): 46–51.

3. For additional analysis, see Timothy W. Luke, *Screens of Power: Ideology, Domination, and Resistance in Informational Society* (Urbana: University of Illinois Press, 1989), 207–39; and Klaus Eder, *The New Politics of Class: Social Movements and Cultural Dynamics in Advanced Societies* (London: Sage, 1993).

4. Bowden, "Dave Foreman!" 46. For a sustained articulation of Earth First! political thinking in connection with deep ecology, see Steve Chase, ed., *Defending the Earth: A Dialogue between Murray Bookchin and Dave Foreman* (Boston: South End Press, 1991); and Bron Raymond Taylor, *Ecological Resistance Movements: The Global Emergence of Radical and Popular Environmentalism* (Albany: State University of New York Press, 1995), 11–34.

5. Carl Boggs, "The Intellectual and Social Movements: Some Reflections on Academic Marxism," *Humanities in Society* 1, nos. 3 and 4 (1983): 228–29.

6. Jean Cohen, "Strategy or Identity: New Theoretical Paradigms and Contemporary Social Movements," *Social Research* 52, no. 4 (1985): 690.

7. Jürgen Habermas, "New Social Movement," *Telos* 49 (fall 1981): 34.

8. Ibid.

9. Ibid., 36.

10. Ibid., 33; emphasis in the original.

11. Joachim Hirsch, "The Fordist Security State and New Social Movement," *Kapitalistate* 10/11 (spring 1984): 84.

12. Ibid.

13. Ibid, 85.

14. See Timothy W. Luke, "Class Contradictions and Social Cleavages in Informationalizing Post-Industrial Sciences: On the Rise of New Social Movements," *New Political Science* 16/17 (1989): 125–53.

15. *Earth First!* November 1, 1987, 20.

16. Ibid.

17. See Michael Parfit, "Earth First!ers Wield a Mean Monkey Wrench," *Smithsonian* 21, no. 1 (April 1990): 186–90.

18. See Zakin, *Coyotes and Town Dogs*, 84–90; and Scarce, *Eco-Warriors*, 91–94.

19. Foreman, *Confessions of an Eco-Warrior*, 55–86.

20. *Earth First! Journal*, May 1, 1990, 3.

21. In states that do not yet have organized groups or in other areas in states with groups, there are another forty-five coordinators. Outside of the United States, Earth First! groups have been organized in Australia, West Germany, Kenya, Mexico, Poland, Scotland (two groups), Canada (three groups), and Sweden (one group). Although Earth First! has no organized oversight or central administration of these various initiatives, a directory of all groups that want to be listed openly as affiliated participants in the movement is maintained by a coordinator in Madison, Wisconsin, and published in every issue of *Earth First! Journal*. Similarly, a large number of the regional and local offices produce their own newsletters or mailings to inform their members or interested others about their diverse activities. See *Earth First!* February 2, 1993, 9.

22. For more discussion, see Martha F. Lee, *Earth First! Environmental Apocalypse* (Syracuse: Syracuse University Press, 1995), 34–36.

23. Zakin, *Coyotes and Town Dogs*, 1–6, 420–43.

24. See ibid., 342–96.

25. *Earth First!* February 2, 1991, 2.

26. Zakin, *Coyotes and Town Dogs*, 397–419.

27. *Wild Earth* (spring 1991): 2.

28. Ibid.

29. Ibid., 6–12.

30. *Earth First!* May 1, 1990, 5.

31. Ibid., 3.

32. "Earth First!" *60 Minutes*, June 3, 1990. After a bandsaw operator was severely injured in May 1987 in a Louisiana Pacific lumber mill outside Clover-

dale, California, tree spiking became an even more highly publicized and controversial practice. Tree spiking has become so threatening that a federal law now makes it punishable with jail terms of up to five years and a thirty thousand dollar fine. The purpose of tree spiking is not to sabotage lumber mills or kill loggers but rather to prevent trees from being cut and milled in the first place. Earth First! rejects violence, yet it also recognizes that violent countermeasures are being directed against environmental, wilderness, and biodiversity activists. Some activists, in turn, such as Paul Watson, the international director of Sea Shepherd, who sees his organization as "the navy to Earth First!'s army," claims that his organization has rammed, sunk, or somehow disabled twelve whaling ships in its defense of marine mammals. See Parfit, "Earth First!ers Wield a Mean Monkey Wrench," 196.

Using violence against ecoactivists is becoming much more common. In the 1980s, for example, gorilla researcher Dian Fossey, Greenpeace photographer Fernando Pereira, and Brazilian rubber tapper Chico Mendes all were killed for their environmental resistance work. Scores have died in Kenya in the war on elephant poachers, and Cultural Survival—the indigenous peoples' rights organization—estimates that more than one person a day dies in the struggle to protect the world's rain forest. See *Earth First!* February 1, 1990, 31. In May 1990, Earth First! activists Judi Bari and Darryl Cherney were injured in a car bombing apparently provoked by their leading role in the "Redwood Summer" campaign being organized in northern California. And, they allegedly have been harassed by many police agencies continuously since 1990. See *Earth First!* June 21, 1994, 14–15.

33. Cited in Bowden, "Dave Foreman!" 49.

34. See Dave Foreman and Howie Wolke, *The Big Outside: A Descriptive Inventory of the Big Wilderness Areas of the USA* (Tucson: Ned Ludd Books, 1989).

35. See *Earth First! Journal*, May 1, 1990, 3–4.

36. For additional discussion, see Dave Foreman, *Ecodefense: A Field Guide to Monkey Wrenching* (Tucson: Ned Ludd Books, 1983).

37. These realities are reflected, in part, but its formal-legal history. The *Earth First! Journal* began in 1980 as a photocopied newsletter assembled by Dave Foreman and Howie Wolke to disseminate news about ecodefense. Its popularity led to adopting a newspaper format and a regular publication schedule that now corresponds to the planet's solstices, equinoxes, and midpoints—or eight times a year. The larger scale of production also made it prudent to become a sole proprietorship business enterprise, which it did under the ownership of Earth First! founding member Dave Foreman. Based in Tucson, Arizona, it was edited, owned, and published by Dave Foreman, who held it as part

of his publishing concern, Ned Ludd Books. Under his independent ownership, however, it was set up to serve the larger Earth First! movement. To keep up with his speaking and writing obligations, Foreman surrendered sole proprietorship of the journal to a small, four-person editorial and managerial collective in January 1989. Foreman also turned over the editorial reins to John Davis, the current editor, in 1988, although he continued to serve as publisher and remained the owner of Ned Ludd Books. Originally part of the journal's operations, Ned Ludd Books published Foreman's *Ecodefense: A Field Guide to Monkey Wrenching* and Foreman and Wolke's *The Big Outside*, an updated version of conservationist Bob Marshall's 1936 inventory of roadless areas. For more discussion, see *Earth First! Journal*, May 1, 1990, 2–5.

38. Ibid., 5.

39. Ibid.

40. See Edward Abbey, *The Monkey Wrench Gang* (Philadelphia: Lippincott, 1975); and *Hayduke Lives!* (New York: Little, Brown, 1989).

41. Cited in Parfit, "Earth First!ers Wield a Mean Monkey Wrench," 194.

42. See Arne Naess, *Ecology, Community, and Lifestyles: Outline of an Ecosophy*, trans. and ed. David Rothenberg (Cambridge: Cambridge University Press, 1989); and Bill Devall and George Sessions, *Deep Ecology* (Salt Lake City: Peregrine Smith Books, 1985).

43. *Earth First! Journal*, May 1, 1990, 4.

44. Ibid., 4–5.

45. Ibid., 5.

46. Brian Tokar, "Social Ecology, Deep Ecology and the Future of Green Political Thought," *Ecologist* 18, nos. 4–5 (1988): 132–41.

47. *Earth First!* November 1, 1987, 20.

48. *Earth First!* February 2, 1990, 25.

49. *Earth First!* November 1, 1987, 20.

50. Ibid.

51. Ibid.

52. *Earth First!* February 2, 1990, 10.

53. *Earth First!* November 1, 1987, 20.

54. Ibid.

55. Ibid. For additional discussion of grizzly bears by Earth First! see *Earth First!* September 23, 1987, 10–11; November 1, 1987, 12; August 1, 1988, 4; November 1, 1988, 13; November 1, 1989, 5; December 21, 1989, 8; and March 20, 1989, 11.

56. *Earth First!* November 1, 1987, 22.

57. Ibid.

58. Ibid., 21.

59. Ibid.

60. Ibid.

61. Ibid.

62. Ibid., 20.

63. Cited in *Earth First!* February 2, 1990, 2.

64. *Earth First!* November 1, 1987, 21.

65. Ibid. For additional defense of monkey wrenching, see *Earth First!* November 1, 1989, 32; December 21, 1989, 30; February 2, 1990, 8; and March 20, 1990, 26.

66. *Earth First!* December 21, 1989, 20–21.

67. Ibid., 20.

68. *Earth First! Journal* May 1, 1990, 4.

69. Even so, Foreman admits that in the early 1980s he and many of his associates were "having fantasies of being the Mahatma Gandhi of the environmental movement" (Zakin, *Coyotes and Town Dogs*, 252).

70. "Earth First!" *60 Minutes*, June 3, 1990.

71. Parfit, "Earth First!ers Wield a Mean Monkey Wrench," 198. Also see Brandon Mitchener, "Out on a Limb for Mother Earth," *E Magazine* 1, no. 1 (January–February 1990), 42–51.

72. *Earth First!* February 2, 1990, 30.

73. Ibid.

74. Ibid.

75. "Earth First!" *60 Minutes*, June 3, 1990.

76. See, for example, the coverage in *Time, Newsweek, U.S. News and World Report, USA Today, New York Times, Los Angeles Times,* or *Washington Post* during 1989 and 1990.

77. *Time* 135, no. 17 (April 23, 1990): 77. In May 1989, the FBI arrested Foreman and four other Earth First!ers for allegedly engaging in a conspiracy to cut the electric supply lines to the Palo Verde Nuclear Generating Station, which is sited west of Phoenix, after three people were caught in the act of cutting on a power pylon with a blowtorch out in the desert. Foreman, however, has been cast as the mastermind and moneybags behind this act of sabotage plus additional strikes allegedly being planned against Diablo Canyon Nuclear Generating Station, Rocky Flats Nuclear Facility, the Central Arizona Project, and a ski resort near Flagstaff, Arizona. After infiltrating Earth First! in Tucson, the FBI put Foreman and the others under surveillance for nearly a year. Believing that Earth First! was indeed a threat to the state, the FBI has put Foreman at

the center of a public relations campaign to represent Earth First!ers as "wild-eyed terrorists of the Palestinian type." See Bowden, "Dave Foreman," 48.

78. *Earth First!* February 2, 1990, 31. Also see *Earth First!* May 1, 1996, 1, 26, for discussion of mainstream media attempts to tie the Unabomber with Earth First!

79. "Earth First!" *60 Minutes*, June 3, 1990.

80. *Earth First!* January 1, 1989, 25.

81. Ibid.

82. See Edward Abbey, *Desert Solitaire: A Season in the Wilderness* (Salt Lake City: Peregrine Smith Books, 1981); *Abbey's Road* (New York: E. P. Dutton, 1981); and *A Fool's Progress* (New York: Henry Holt, 1988).

83. See Parfit, "Earth First!ers Wield a Mean Monkey Wrench," 188. For more discussion of these issues of political and media styles, see Zakin, *Coyotes and Town Dogs*, 170–71, 271–305. As Taylor observes, Earth First! "is pluralistic and constantly changing," but the "buckaroo persona" that has been mapped onto the group is much more than a misreading from old magazine articles (see *Ecological Resistance Movements*, 11–15).

84. Bowden, "Dave Foreman!" 48 and 51.

85. *Earth First!* February 2, 1990, 30.

86. Ibid.

87. Bowden, "Dave Foreman!" 51.

88. *Earth First!* March 20, 1990, 26.

89. "Earth First!" *60 Minutes*, June 3, 1990.

90. See Paul Wapner, *Environmental Activism and World Civic Politics* (Albany: State University of New York Press, 1996), 54–71, for a discussion of "stinging" tactics in environmental politics. A good sense of Earth First!'s multinational following can be had in *Earth First!* May 1, 1996, which details actions in Canada, the United States, Russia, the United Kingdom, Japan, and the Navajo nation's reservation in the American Southwest.

91. See Hazel Henderson, *Creating Alternative Futures: The End of Economics* (New York: Berkeley, 1978).

92. See Timothy W. Luke, "Informationalism and Ecology," *Telos* 56 (summer 1983): 59–73.

3. The Nature Conservancy or the Nature Cemetery:
Buying and Selling "Perpetual Care" as Environmental Resistance

1. See Noel Grove, "Quietly Preserving Nature," *National Geographic* 174, no. 6 (December 1988): 834.

2. Noel Grove, *Preserving Eden: The Nature Conservancy* (New York: Henry Abrams, 1988), 25.

3. Ibid.

4. The Nature Conservancy, "Dear Investor," a direct mail membership package (Arlington, Va.: The Nature Conservancy, 1994), 1–2.

5. Ibid.

6. Ibid.

7. Ibid.

8. Ibid.

9. Ibid., 2.

10. Ibid., 3–4.

11. Ibid., 3.

12. Ibid.

13. Ibid.

14. Ibid., 1.

15. Ibid.

16. Ibid., 4.

17. Ibid., 2.

18. Ibid., 3.

19. Ibid., 2.

20. *Virginia Chapter News: Summer 1994* (Charlottesville, Va.: The Nature Conservancy, 1994), 8 and 9.

21. Ibid., 9.

22. Ibid., 11.

23. Ibid.

24. Ibid., 4.

25. Ibid.

26. Ibid.

27. For a sense of this reading of Nature in The Nature Conservancy philosophy, see Max Oelschlaeger, *The Idea of Wilderness: From Prehistory of the Age of Ecology* (New Haven: Yale University Press, 1991).

28. Bill McKibben, *The End of Nature* (New York: Doubleday, 1989); and Carolyn Merchant, *The Death of Nature: Women and the Scientific Revolution* (San Francisco: Harper and Row, 1990).

29. See Timothy W. Luke, "On Environmentality: Geo-Power and Eco-Knowledge in the Discourses of Contemporary Environmentalism," *Cultural Critique* 31 (fall 1995): 57–81.

30. Grove, "Quietly Preserving Nature," 837.

31. For an extended elaboration of "biodiversity" as a concept and prac-

tice, see E. O. Wilson, ed., and Frances M. Peter, associate ed., *Biodiversity* (Washington, D.C.: National Academy Press, 1988).

32. Grove, *Preserving Eden*, 30–31.

33. *Virginia Chapter News: Summer 1994*, 2.

34. Ibid.

35. Ibid.

36. Ibid.

37. Ibid., 1.

38. Ibid.

39. Ibid.

40. The Nature Conservancy, "Dear Investor," 1–3.

41. Ibid., 2.

42. Ibid.

43. Ibid. To extend these nice feelings at the shopping mall, the 1996 package of membership privileges also entitles Nature Conservancy members to "a 10 percent discount on most purchases from the Nature Company."

44. Grove, *Preserving Nature*, 35.

45. Nearly 25 percent of all drugs come from plants, but less than 1 percent of all natural plants have been studied for their pharmaceutical potential. As Michael J. Balick of the New York Botanical Garden notes, there has been an explosion of interest in plants to develop new drugs; see Kathleen Day, "Remedies from the Rain Forest," *Washington Post* (September 19, 1995), E1, 4. For a sense of the economic potentialities here, see Curtis Moore and Alan Miller, *Green Gold: Japan, Germany, the United States and the Race for Environmental Technology* (Boston: Beacon Press, 1994).

46. Another consideration of this construction of Nature can be found in George Robertson et al., *Future Natural: Nature/Science/Culture* (London: Routledge, 1996), 107–45.

4. Worldwatching at the Limits of Growth

1. Lester Brown et al., *State of the World* (New York: Norton, 1991), vii.

2. Ibid.

3. Lester Brown, Christopher Flavin, and Sandra Postel, *Saving the Planet: How to Shape an Environmentally Sustainable Society* (New York: Norton, 1991), dust jacket.

4. Ibid.

5. Ibid.

6. Ibid., 21.

7. Ibid., 22.

8. Ibid.

9. Ibid., 23.

10. Ibid.

11. For additional discussion, see David Noble, *America by Design: Science, Technology, and the Rise of Corporate Capitalism* (New York: Knopf, 1977).

12. Henry Jarrett, ed., *Perspectives on Conservation: Essays on America's Natural Resources* (Baltimore: Resources for the Future, 1958), 51.

13. See Donald Worster, *Nature's Economy: The Roots of Ecology* (Garden City, N.Y.: Anchor Books, 1977), for a more elaborate analysis of these issues.

14. Brown, Flavin, and Postel, *Saving the Planet,* 73.

15. Ibid., 73–74.

16. Ibid., 74.

17. Ibid., 31.

18. Ibid., 32.

19. On this point, see Amory Lovins, *Soft Energy Paths* (New York: Ballinger, 1977); and David Dickson, *Alternative Technology and the Politics of Technical Change* (Glasgow: Fontana, 1974).

20. This sort of micrological household-centered "worldwatching" can be seen in works like Debra Dadd-Redalia, *Sustaining the Earth: Choosing Consumer Products That Are Safe for You, Your Family, and the Earth* (New York: Hearst Books, 1994).

21. Brown, Flavin, and Postel, *Saving the Planet,* 173.

22. Ibid.

23. Ibid., 179–80.

24. Ibid., 179.

25. See, for example, Michael Redclift, *Sustainable Development: Exploring the Contradictions* (London: Methuen, 1987); and Robert Riddell, *Ecodevelopment* (London: Gower, 1981).

26. World Commission on Environment and Development, *Our Common Future* (New York: Oxford University Press, 1987), 46.

27. Brown, Flavin, and Postel, *Saving the Planet,* 179, 180.

28. Ibid., 180.

29. Donella Meadows et al., *The Limits to Growth* (New York: Unicorn Books, 1972).

30. Ibid., 27.

31. Ibid., xv.

32. Redclift, *Sustainable Development,* 36.

33. Lester Brown, *Building a Sustainable Society* (New York: Norton, 1981), 247.

34. Ibid., 248.

35. Ibid., 248–83.

36. Ibid., 285.

37. Ibid., 311.

38. Ibid., 314.

39. Ibid., 317.

40. Ibid., 322.

41. Ibid., 323.

42. Ibid., 324.

43. Ibid., 325.

44. Ibid., 326.

45. For an elaboration of these notions, see Warren Johnson, *Muddling toward Frugality: A Blueprint for Survival in the 1980s* (Boulder, Colo.: Shambhala, 1978); E. F. Schumacher, *Small Is Beautiful: Economics as if People Mattered* (New York: Harper and Row, 1973).

46. Fredric Jameson, *Postmodernism, or the Cultural Logic of Late Capitalism* (Durham, N.C.: Duke University Press, 1991), ix.

47. See Michel Foucault, "Governmentality," in *The Foucault Effect: Studies in Governmentality*, ed. Graham Burchell, Colin Gordon, and Peter Miller (Chicago: University of Chicago Press, 1991), 93.

48. Michel Foucault, *The History of Sexuality*, vol. 1, *An Introduction* (New York: Vintage, 1960), 138–42. The statistical surveillance regime of states, Foucault maintains, emerges alongside monarchical absolutism during the late seventeenth century. Intellectual disciplines, ranging from geography and cartography to statistics and civil engineering, are mobilized to inventory and organize the wealth of populations in territories by the state. For additional discussion, see Burchell, Gordon, and Miller, eds., *The Foucault Effect*, 1–48.

49. Ibid., 143.

50. Ibid., 142.

51. Ibid.

52. Lester Brown et al., *State of the World* (New York: Norton, 1988), xv.

53. Ibid.

54. Ibid., 321. This metaphor, in turn, has led to a new, shorter survey of global ecological trends titled *Vital Signs: The Trends That Are Shaping Our Future* (New York: Norton, 1992–96), to "chart a sustainable future."

55. Brown et al., *State of the World* (1988), 21.

56. Brown, Flavin, and Postel, *Saving the Planet*, 25.

57. The aura of knowledge offered by such systems of global surveillance moves many to engineer new "social-choice mechanisms" in new "designs and

metadesigns" capable of remaking all existing social and political institutions into practical enhancements of ecological rationality. For some articulation of these tendencies, see John S. Dryzek, *Rational Ecology: Environment and Political Economy* (Oxford: Blackwell, 1987); and John O'Neill, *Ecology, Policy and Politics: Human Well-Being and the Natural World* (London: Routledge, 1993). Additional consideration of the socially constructed qualities of environmental knowledge can be found in John A. Hannigan, *Environmental Sociology: A Social Constructionist Perspective* (London: Routledge, 1995).

5. Environmental Emulations: Terraforming Technologies and the Tourist Trade at Biosphere 2

1. For more discussion, see William J. Broad, "Troubled Biosphere Project Gets Help from Pair in the Ivy League," *New York Times* (August 14, 1993), 20; and David L. Wheeler, "New Lease for Biosphere 2," *Chronicle for Higher Education*, 41, no. 2 (September 7, 1993), A12, A27. On January 1, 1996, Columbia University assumed responsibility for managing all aspects of Biosphere 2, and it set about turning the facility into a unique new educational center and environmental test bed. See Jonathan Rabinovitz, "Columbia to Take Over Biosphere 2 as Earth Lab," *New York Times* (November 13, 1995), B2.

2. See Robert Lee Hotz and Adam S. Bauman, "Biosphere 2: Trouble in Paradise," *Los Angeles Times* (April 24, 1994), A1; B. Drummond Ayres Jr., "Ecological Experiment Becomes Battle Ground: Biosphere 2," *New York Times* (April 11, 1994), A9; Adam S. Bauman, "Creator of Biosphere 2 Quits Project: John P. Allen," *Los Angeles Times* (April 9, 1994); Don Phillips, "Biosphere 2 Managers Ousted in Raid," *Washington Post* (April 3, 1994), A5; Robert S. Hotz and Adam S. Bauman, "Key Investor's Agents Seize Biosphere 2: Edward P. Bass in Management Dispute," *Los Angeles Times* (April 2, 1994), A1; and Colin Macilwain, "Palace Coup May Lift Status of Biosphere 2," *Nature* 368 (April 14, 1994): 576.

3. John Allen, *Biosphere 2: The Human Experiment* (New York: Penguin, 1991), 18.

4. Ibid., 18–19.

5. Ibid., 3.

6. For discussions of the Biosphere's inaugural mission, see Craig Kasnoff, "Live-in Lab Is Its Own World; Biosphere 2 Creates Environment for Studying Global Processes and Mankind's Impact on Them," *Christian Science Monitor* (February 20, 1991), 12; Seth Mydans, "8 Seek Better World in 2-Year Ecology Project: Biosphere 2," *New York Times* (September 27, 1991), A1; and "Little Big World: Biosphere 2," *Economist* 320 (September 21, 1991): 106.

7. Reports of these setbacks can be found in Ian Mundell, "Biosphere 2 Project Told to Make Room for Science," *Nature* 358 (August 6, 1992): 444; Joel Achenbach, "Biosphere 2: Bogus New World?" *Washington Post* (January 8, 1992), C1; Edie Jarolim, "Yuppie Utopia: Touring the Biosphere," *Wall Street Journal* (May 26, 1992), A14; Traci Watson, "Can Basic Research Find a Good Home in Biosphere 2," *Science* 259 (March 19, 1993): 1688; Sharon Begley, "New Cracks in the Glass House: Science Advisors Resign from Biosphere 2," *Newsweek* 121 (March 1, 1993): 67; William J. Broad, "Oxygen Loss Causing Concern in Biosphere 2; Scientists Mystified by 28% Drop in Miniature World," *New York Times* (January 5, 1993), C4; Seth Mydans, "8 Bid Farewell to the 'Future': Musty Air, Roaches and Ants; Biosphere 2 Inhabitants Leave Experimental Self-Contained Dome," *New York Times* (September 27, 1993), A1; and Philip Elmer-Dewitt, "Getting Back to Earth: Biosphere 2," *Time* 142 (October 4, 1993): 91.

8. Wheeler, "New Lease for Biosphere 2," A12. Since September 1994, the new scientific advisory panel has shifted the focus away from humans living and working in Biosphere 2 and to the earth itself, using the structure as a test module for understanding changes in Biosphere 1. Humans use the structure, but they do not inhabit it. Indeed, the entire human biome was simply shut down after its human and domesticated animal populations were removed when the second mission ended. Several months into its redirection toward "big science," Bannon assured the Tucson media about the more serious scientific import of the new team's work: "we are exclusively dedicated to science and to education. Over the next few years, you will see Biosphere 2 evolve into a leading, if not the leading, center for the study of global systems. What the particle accelerator was going to do for Texas, Biosphere 2 has the potential for doing right here" (*Tucson Citizen* [February 18, 1995], 1–B). While likening Biosphere 2 to the failed Texas supercollider may not be good public relations, it did underscore Bannon's interpretation of his administration of the facility: "We essentially are hitting the reset button" in a series of operational steps that "will clear the decks for the first phase of a long-term science plan and establish a base-line for all future experiments" (*The Biosphere: The Magazine of the Biosphere 2 Project* 2, no. 1 [spring 1995]: 19).

9. Biosphere 2, *Visitor Map and Directory* (Oracle, Ariz.: Space Biospheres Ventures, 1991).

10. Linnea Gentry and Karen Liptak, *The Glass Ark: The Story of Biosphere 2* (New York: Puffin Books, 1991), 3–20.

11. Ibid., 21–35.

12. Biosphere 2, in fact, consumes immense amounts of energy and labor

simply to stay up and running. See Tim Appenzeller, "Biosphere 2 Makes a New Bid for Scientific Credibility," *Science* 263, no. 5152 (March 11, 1994): 1368–70; and Tim Beardsley, "Down to Earth: Biosphere 2 Tries to Get Real," *Scientific American* 273, no. 2 (August 1995): 24–26.

13. The solar and atmospheric efficiencies of Biosphere 2 are a bit problematic. Its glass skin and metal space frames restrict sunlight by 50 percent, but only 10 percent of its atmosphere is lost every year. See Beardsley, "Down to Earth," 24–25.

14. Although it started to simulate an "earth environment" for human beings, this environmental pastiche is being readapted to host more serious scientific studies of community ecology, sustainable agriculture, and plant physiology (ibid).

15. Mark Nelson and Gerald Soffen, *Biological Life Support Systems* (Oracle, Ariz.: Synergetic Press, 1990), vii.

16. Ibid., vii–viii.

17. Biosphere 2, *Visitor Map and Directory*.

18. Ibid.

19. Wheeler, "New Lease for Biosphere 2," A12.

20. Nelson and Soffen, *Biological Life Support Systems*, 60.

21. Ibid., 66.

22. See John Allen and Mark Nelson, *Space Biospheres* (Oracle, Ariz.: Synergetic Press, 1989), ii.

23. See Dorion Sagan, *Biospheres: Reproducing Planet Earth* (New York: Bantam Books, 1990), 3–11.

24. Allen and Nelson, *Space Biospheres*, 1.

25. Ibid., 75.

26. Vladimir Vernadsky, *The Biosphere* (Oracle, Ariz.: Synergetic Press, 1986), 11.

27. Allen, *Biosphere 2*, 12–13.

28. Sagan, *Biospheres*.

29. Allen and Nelson, *Space Biospheres*, 52.

30. Ibid., 40.

31. See Kevin Kelly, *Out of Control: The Rise of New Biological Civilization* (Reading, Mass.: Addison-Wesley, 1994), 147.

32. Ibid.

33. Allen and Nelson, *Space Biospheres*, 58.

34. Ibid. For another reconsideration of Biosphere 2's design quirks from an original team member, see Peter Warshall, "Lessons from Biosphere 2: Ecodesign, Surprises, and the Humility of Gaian Thought," *Whole Earth Review* 89

(spring 1996): 22–27. For additional information on the travails of the Biosphere 2 structure and its human inhabitants during the 1991–94 experimental missions, see Gabrielle Walker, "Secrets from Another Earth," *New Scientist* 151 (May 18, 1996): 31–35; and Joel E. Cohen and David Tilman, "Biosphere 2 and Biodiversity: The Lessons So Far," *Science* 274 (November 15, 1996): 1150–51. The basic wrongheadedness of the initial Biosphere 2 experiments is summarized succinctly by Cohen and Tilman: "Despite the enormous resources invested in the original design and construction (estimated at roughly $200 million from 1984 to 1991) and despite a multi-million dollar operating budget, it proved impossible to create a materially closed system that could support eight human beings with adequate food, water, and air for two years. Isolating small pieces of large biomes and juxtaposing them in an artificial enclosure changed their functioning and interactions rather than creating a small working Earth" (1150).

35. "The Big Picture—Columbia University's Global Systems Initiative: Rethinking the Way People Think," *The Biosphere: The Magazine of the Biosphere 2 Project* 2, no. 1 (spring 1995): 11.

36. Ibid.

6. Green Consumerism: Ecology and the Ruse of Recycling

1. See, for example, Barry Commoner, *Science and Survival* (New York: Viking, 1966); and Murray Bookchin, *Postscarcity Anarchism* (Berkeley: Ramparts Books, 1971).

2. See John G. Mitchell and Constance L. Stallings, eds., *Ecotactics: The Sierra Club Handbook for Environmental Activists* (New York: Pocket Books, 1970), 13–35.

3. This green engagement of big business, in turn, often is seconded by theorists of mass consumption who see large numbers of consumers steering capital toward more ecological forms of life through their buying behavior in shopping decisions. See, for example, Mica Nava, "Consumerism Reconsidered: Buying and Power," *Critical Studies* 5 (May 1991): 157–72.

4. See Samuel Hays, *Beauty, Health, and Permanence: Environmental Politics in the United States, 1955–1985* (Cambridge: Cambridge University Press, 1987); and Frank E. Smith, *The Politics of Conservation* (New York: Pantheon, 1966).

5. Hays, *Beauty, Health, and Permanence*, 137–245.

6. Heloise, *Hints for a Healthy Planet* (New York: Perigee Books, 1990).

7. Earth Works Group, *50 Simple Things You Can Do to Save the Earth* (Berkeley: Earthworks Press, 1991), 6. From 1988 to 1996, the Environmental Defense Fund reports that curbside recycling increased tenfold, and 125 million

Americans now have some kind of curbside recycling service. As a result, polyethylene terephthalate (PET) plastic bottles in 1994 had a 48 percent recycling rate, and 62 percent of all aluminum cans in 1995 were recycled, according to trade association figures. See Alexandra Robbins, "Use It Once, Use It Again: The Recycling Circle Makes Sense All Around," *Washington Post* (June 17, 1996), C5.

8. Ibid.

9. Ibid.

10. Ibid., 7.

11. Ibid., 6.

12. Ibid., 7.

13. Ibid., 18.

14. Earth Works Group, *The Recycler's Handbook: Simple Things You Can Do* (Berkeley: Earthworks Press, 1990).

15. Earth Works Group, *The Next Step: 50 More Things You Can Do to Save the Earth* (Kansas City: Andrews and McMeel, 1991), 7.

16. Robert D. Holsworth, *Public Interest Liberalism and the Crisis of Affluence: Reflections on Nader, Environmentalism, and the Politics of a Sustainable Society* (Boston: G. K. Hall, 1980), 28–72.

17. Earth Works Group, *The Student Environmental Action Guide: 25 Simple Things We Can Do* (Berkeley: Earthworks Press, 1991).

18. The Earth Works Group, *50 Simple Things Your Business Can Do to Save the Earth* (Berkeley: Earthworks Press, 1991).

19. Ibid., 9.

20. Ibid., 8.

21. Ibid.

22. Ibid., 9.

23. Jeremy Rifkin, ed., *The Green Lifestyle Handbook: 1001 Ways You Can Heal the Earth* (New York: Henry Holt, 1990).

24. Ibid., 130.

25. Ibid.

26. Holsworth, *Public Interest Liberalism*, 28–72.

27. Jeffrey Hollander, *How to Make the World a Better Place: A Guide to Doing Good* (New York: William Morrow, 1990), 9.

28. Ibid.

29. Ibid., 19.

30. Ibid., 21.

31. Ibid.

32. Ibid.

33. Ibid.

34. Ibid.

35. Ibid., 31.

36. Ibid.

37. Ibid., 33.

38. Marjorie Lamb, *Two Minutes a Day for a Greener Planet: Quick and Simple Things You Can Do to Save Our Earth* (New York: HarperCollins, 1990).

39. Ibid., i.

40. *The St. Louis Post-Dispatch* (July 8, 1991), 1BP.

41. Ibid.

42. Ibid., 8BP.

43. *Consumer Reports* (October 1991): 687.

44. Ibid., 689.

45. Ibid.

46. *New York Times* (April 22, 1990), 24.

47. Ibid.

48. *The Roanoke Times & World News* (April 22, 1990), 5.

49. *New York Times* (April 22, 1990), 24.

50. Ibid., 25.

51. Ibid.

52. For another revisionist reading of recycling, see John Tierney, "Recycling Is Garbage," *New York Times Magazine* (June 30, 1996), 24–29, 44, 48, 51, 53. This discussion sees recycling as a mostly inefficient response to green evangelists, who have turned recycling into a virtuous response to excessive consumption. Yet, it also increased waste disposal costs, added new government regulations, failed to reduce industrial pollutants substantially, and burdened consumers with new household chores, although some evidence shows that it does save energy and reduce some types of air and water pollution.

7. Marcuse and the Politics of Radical Ecology

1. See Herbert Marcuse, *Counterrevolution and Revolt* (Boston: Beacon Press, 1972); and *The Aesthetic Dimension* (Boston: Beacon Press, 1978).

2. Langdon Winner, *The Whale and the Reactor: A Search for Limits in an Age of High Technology* (Chicago: University of Chicago Press, 1982), 69. See also Hazel Henderson, *The Politics of the Solar Age* (Garden City, N.Y.: Anchor Press, 1981), 167–68. Robyn Eckersley picks up this theoretical trail with her critique of the Frankfurt School in *Environmentalism and Political Theory* (Albany: State University of New York Press, 1992), 97–106.

3. Koula Mellos, *Perspectives on Ecology* (New York: St. Martin's Press, 1988), 4.

4. For examples of these readings, see Murray Bookchin, *The Ecology*

of Freedom (Palo Alto, Calif.: Cheshire Books, 1982); and Christopher Manes, *Green Rage: Radical Environmentalism and the Unmaking of Civilization* (Boston: Little, Brown, 1990).

5. See Marcuse, *Counterrevolution and Revolt*, 59–78.

6. Herbert Marcuse, "Ecology and Revolution: A Symposium," *Liberation* 17, no. 6 (1972).

7. Herbert Marcuse, "Ecology and the Critique of Modern Society," *Capitalism Nature Socialism* 3, no. 3 (September 1992): 29–38.

8. Morton Schoolman, *The Imaginary Witness* (New York: Free Press, 1980); and Douglas Kellner, *Herbert Marcuse and the Crisis of Marxism* (Berkeley: University of California Press, 1984).

9. See Herbert Marcuse, *Eros and Civilization* (Boston: Beacon Press, 1955); *One Dimensional Man* (Boston: Beacon Press, 1964); and *An Essay on Liberation* (Boston: Beacon Press, 1969).

10. Marcuse, *Counterrevolution and Revolt*, 60.

11. Ibid., 59.

12. Marcuse, *One Dimensional Man*, 3.

13. Ibid., 3.

14. Ibid., 4.

15. Marcuse, *Eros and Civilization*, 225.

16. Marcuse, *One Dimensional Man*, 2.

17. Ibid., 5.

18. Ibid., 4–5.

19. Ibid., 5.

20. Ibid., 7.

21. Ibid.

22. Marcuse, "Ecology and the Critique of Modern Society," 36.

23. Ibid.

24. Ibid., 32.

25. Ibid., 36.

26. See Max Horkheimer and T. W. Adorno, *Dialectic of Enlightenment* (New York: Seabury, 1972).

27. Marcuse, *One Dimensional Man*, 157–58.

28. Ibid., 158.

29. Ibid., 159.

30. Herbert Marcuse, *Five Lectures: Psychoanalysis, Politics, and Utopia* (Boston: Beacon Press, 1970), 12.

31. Marcuse, *One Dimensional Man*, 166.

32. Ibid., 230. For another defense of Marcuse's views of technological

rationality, see William Leiss, *The Domination of Nature* (Boston: Beacon Press, 1974), 199–212.

33. Marcuse, *An Essay on Liberation*, 31.

34. Ibid., 31. See also Marcuse, *The Aesthetic Dimension*, 54–69.

35. Ibid., 31–32.

36. See Marcuse, *Counterrevolution and Revolt*, 59–78.

37. Marcuse, *Eros and Civilization*, 131.

38. Ibid.

39. Marcuse, *Counterrevolution and Revolt*, 74.

40. Ibid., 75.

41. Ibid., 74.

42. Marcuse, *One Dimensional Man*, 240.

43. Ibid., 236.

44. Ibid., 242.

45. Ibid.

46. Ibid., 252–53.

47. Ibid., 247.

48. See Barry Commoner, *Science and Survival* (New York: Viking, 1963); *The Closing Circle: Nature, Man and Technology* (New York: Knopf, 1971); *The Poverty of Power* (New York: Knopf, 1976); and Murray Bookchin, *Post-Scarcity Anarchism* (Berkeley: Ramparts Press, 1971).

49. See Rachel Carson, *Silent Spring* (Boston: Houghton, Mifflin, 1964); Herman Daly, *Toward a Steady-State Economy* (San Francisco: W. H. Freeman, 1973); and David Brower, *Not Man Apart* (San Francisco: Sierra Club Books, 1965).

50. See Arne Naess, *Community, Ecology, and Lifestyle* (Cambridge: Cambridge University Press, 1989); Bill Devall and George Sessions, *Deep Ecology* (Salt Lake City: Peregrine Smith Books, 1985); Bill Devall, *Simple in Means, Rich in Ends: Practicing Deep Ecology* (Salt Lake City: Peregrine Smith Books, 1988); George Sessions, "Shallow and Deep Ecology: A Review," in *Ecological Consciousness: Essays from the Earthday X Colloquium* (Washington, D.C.: University Press of America, 1981); E. F. Schumacher, *Small Is Beautiful: Economics as if People Mattered* (New York: Harper and Row, 1973); Dave Foreman, *Confessions of an Eco-Warrior* (New York: Harmony Books, 1991); Ivan Illich, *Energy and Equity* (New York: Harper and Row, 1974); Thomas Berry, *The Dream of the Earth* (San Francisco: Sierra Club Books, 1989); Carolyn Merchant, *The Death of Nature: Women, Ecology and the Scientific Revolution* (New York: Harper and Row, 1980); Henryk Skolimowski, *Living Philosophy: Eco-Philosophy as a Tree of Life* (London: Penguin, 1992); Wendell Berry, *The Unsettling of America:*

Culture and Agriculture (New York: Avon Books, 1977); Wendell Berry, *Standing by Words* (San Francisco: North Point Press, 1983); Wendell Berry, *What Are People For?* (San Francisco: North Point Press, 1990); Bill McKibben, *The End of Nature* (New York: Random House, 1989); Kirkpatrick Sale, *Human Scale* (New York: Coward, McCann and Geoghegan, 1980); and Kirkpatrick Sale, *Dwellers in the Land: The Bioregional Vision* (Philadelphia: New Society Press, 1991).

51. See Roderick Nash, *Wilderness and the American Mind*, 3d ed. (New Haven: Yale University Press, 1982); Samuel Hays, *Beauty, Health, and Permanence: Environmental Politics in the United States* (Cambridge: Cambridge University Press, 1987); and Anna Bramwell, *Ecology in the 20th Century* (New Haven: Yale University Press, 1989).

52. See Donald Edward Davis, *Ecophilosophy: A Field Guide to the Literature* (San Pedro, Calif.: R. and E. Miles, 1989).

53. See Timothy W. Luke, "A Phenomenological/Freudian Marxism? Marcuse's Critique of Advanced Industrial Society," in *Social Theory and Modernity: Critique, Dissent, and Revolution* (Newbury Park, Calif.: Sage, 1990), 128–58.

54. Marcuse, *Eros and Civilization*, 130.

55. For a sense of the potential behind reimagining the built and natural environment in pacified terms, see Paul Hawken, *Ecology of Commerce: A Declaration of Sustainability* (New York: HarperCollins, 1993); Victor Papanek, *The Green Imperative: Ecology and Ethics in Design and Architecture* (London: Thames and Hudson, 1995); and David Wann, *Deep Design: Pathways to a Livable Future* (Washington, D.C.: Island Press, 1996).

56. See, for example, Lester R. Brown, *Building a Sustainable Society* (New York: Norton, 1981); Lester R. Brown, Christopher Flavin, and Sandra Postel, *Saving the Planet: How to Shape an Environmentally Sustainable Economy* (New York: Norton, 1991); Garrett Hardin, *The Limits of Altruism* (Bloomington: Indiana University Press, 1977); and Garrett Hardin, *Filters against Folly: How to Survive Despite Economists, Ecologists, and the Merely Eloquent* (New York: Viking, 1985).

8. Developing an Arcological Politics: Paolo Soleri on Ecology, Architecture, and Society

1. See Paolo Soleri, *Arcosanti: An Urban Laboratory?* (Santa Monica, Calif.: VTI Press, 1987); and Ralph Blumenthal, "Futuristic Visions in the Desert," *New York Times* (February 1, 1987).

2. For more discussion, see Paolo Soleri, "Flight from Flatness," in *The*

Bridge, between Matter and Spirit Is Matter Becoming Spirit: The Arcology of Paolo Soleri (Garden City, N.Y.: Anchor Books, 1973), 198–201.

3. Soleri uses the "arcology" term in many ways. It can be used to describe a method of thinking, the practice of design, an approach to environmental organization, or an actual physical structure that embodies its principles. For more analysis, see Luca Zevi, "Paolo Soleri: A Message to Be Dug Out," *L'architettura: cronache e storia* 422 (December 1990): 849–50.

4. See Paolo Soleri, *Arcology: Architecture in the Image of Man* (Cambridge: MIT Press, 1973).

5. See Francesco Ranocchi, "From a House to a Piece of City: Paolo Soleri's Ideas and Experiments," *L'Architettura: cronache e storia* 422 (December 1990): 856–57.

6. Paolo Soleri and Scott M. Davis, *Paolo Soleri's Earthcasting for Sculpture, Models and Construction* (Salt Lake City: Peregrine Smith Books, 1984), 2.

7. Ibid., 4.

8. For good examples of this quality, see Paolo Soleri, *The Omega Seed: An Eschatological Hypothesis* (Garden City, N.Y.: Anchor Books, 1981).

9. Ibid., 206.

10. Soleri, *The Omega Seed,* 217.

11. Ibid., 223.

12. Ibid., 225.

13. Ibid., 231–32.

14. Paolo Soleri, *Technology and Cosmogenesis* (New York: Paragon House, 1985), viii.

15. For more elaboration of the philosophies advanced by Teilhard de Chardin, see Teilhard de Chardin, *The Phenomenon of Man* (New York: Harper and Row, 1961); *Building the Earth* (Wilkes-Barre, Pa.: Dimension Books, 1965); *The Appearance of Man* (New York: Harper and Row, 1965); *Hymn of the Universe* (New York: Harper and Row, 1965); and *Man's Place in Nature: The Human Zoological Group* (New York: Harper and Row, 1965). Additional discussion of this Teilhardian theology/technology fusion can be found in James Edwards Carlos, "Community of Souls: Soleri/Teilhard; Arcology/Theory," *St. Luke's Journal of Theology* 14 (1971): 134–42; Henryk Skolimowski, "Paolo Soleri: The Philosophy of Urban Life," *Architectural Association Quarterly* 3 (1971): 134–42; Henryk Skolimowski, "Teilhard, Soleri, and Evolution," *Eco-Logos* 22 (1976): 793–800; Russell Lewis, "Residual Anguish, Compassion and Aesthetogenesis in Soleri's Arcologies," *Teilhard Review* 12 (1977): 44–47; and Henryk Skolimowski, "Arcology as an Expression of Process Theology," *Teilhard Review* 12 (1977): 238–39.

16. Soleri, *Technology and Cosmogenesis*, 120–21.

17. Paolo Soleri, *Fragments: A Selection from the Sketch Books of Paolo Soleri* (New York: Harper and Row, 1981), 190.

18. Soleri, *Arcosanti: An Urban Laboratory?* 5.

19. Ibid.

20. Ibid.

21. Ibid., 38. Humanity, or "wo-man" in Soleri's term, is the outcome of three billion years of life's evolution on Earth. Life's creation is continuously generating new life-forms, struggling to realize new existences for conscious, sensitive, creative beings such as human beings. The thinking and doing of such conscious beings are sources of spiritual genesis: "The premise that *homo faber* is also a symbol of *cosmos faber*, the 'fabricating' the cosmos is engaged in, the cosmo-genesis that providentially is a soul genesis. The genesis of spirit is not of spirit (God) generating matter with all the ultimately inconsequential history of a dispensable process of being; it is instead matter generating spirit within the excruciatingly noble and suffering process of metamorphosis. Therefore the urgency of *doing as thinking*. Because the thinking per se—and I include in it contemplating-mediating—is the beautiful flower of a *doing* that has generated brain-mind, that 'thinking machine-performance' of immense nobility and resourcefulness at the apex of the pyramid (hierarchy) of the food chain. Without such a monumental pyramid, thinking-mediating-transcending is impossible, in fact, inconceivable" (Soleri, *Technology and Cosmogenesis*, 133). As with other idealist philosophies of history, ranging from Augustine to Hegel to Teilhard, humanity is the critical mediation of the being/becoming dialectic. Soleri gives humanity primacy, because humans, as far as he can see, sit atop the global food chain and consciousness continuum.

22. Ibid., 19.

23. Ibid., 20.

24. Soleri, *Arcosanti: An Urban Laboratory?* 92.

25. Soleri, *Technology and Cosmogenesis*, 139.

26. Ibid., 137–38.

27. Ibid., 139.

28. Soleri, *Arcology: Architecture in the Image of Man*, 122.

29. Ibid., 14.

30. Ibid., 36–61.

31. Ibid., 62–118.

32. See Tim Luke, "Informationalism and Ecology," *Telos* 56 (summer 1983): 59–73.

33. For a mix of these kinds of readings, see G. Dennis, "Arcosanti: A New

Dream Out on the Desert," *Technology Review* 81 (1979): 16–21; Peter Plagens, "A Visit to Soleri's El Dorado," *Art in America* 67 (1979): 65–71; Jerome C. Glenn, "Prototype Communities of Tomorrow: Arcosanti," *The Futurist* 14 (1980): 35–43; J. Tevere McFadyen, "The Abbot of Arcosanti," *Horizon* 23 (1980): 54–61; M. Basil Pennington, "Arcosanti Monastery No. 1," *America* 144 (March 14, 1981): 207–9; "A Dream City or Ghost Town?" *Newsweek* 97 (March 23, 1981): 14; Naomi M. Bloom, "Human Beehives: Paolo Soleri's Arcosanti," *Science Digest* (March 1981): 42–47; Martin Grosswirth, "Arcosanti: A Laboratory for Living," *SciQuest* 54 (1981): 11–15; Susan Hazen-Hammond, "Mecca or Mirage?" *Discovery* (summer 1985); "A Pilgrimage to Arcosanti," *Arthur Frommer's New World of Travel* (1988); Paul Weingarten, "Futuristic City a Radical Vision Still Out of Focus," *Chicago Tribune* (July 10, 1988); Leonard David, "Paolo Soleri: Man for All Seasons," *Ad Astra* 1 (November 1989): 31; Jane Dodds, "Paolo Soleri," *Art in America* 78 (November 1990): 201; John Poppy, "Home, Sweet Home, 2001-Style," *Longevity* (May 1991): 50–60; "Paolo Soleri's Arcology: Updating the Prognosis," *Progressive Architecture* 72 (March 1991): 76–78; Mark Pastin, "For Selfish Reasons, Arizonans Should Look Again to Arcosanti," *Business Journal* (May 20, 1991); David W. Dunlap, "Future Metropolis," *Omni* 7 (October 1984): 116–24.

34. For more discussion, see Mike Davis, *City of Quartz: Excavating the Future in Los Angeles* (New York: Vintage, 1992); Edward Soja, *Postmodern Geographies: The Reassertion of Space in Critical Social Theory* (Oxford: Blackwell, 1989); Bradford Luckingham, *Phoenix: The History of a Southwestern Metropolis* (Tucson: University of Arizona Press, 1989); and Peter Wiley and Robert Gottlieb, *Empires in the Sun: The Rise of the New American West* (Tucson: University of Arizona Press, 1982).

35. The monastic grounding in Soleri's thinking is quite strong. In *Technology and Cosmogenesis*, he asserts: "One of the paradoxes of spirituality is that it is bound to (generated by) materiality and there, at the critical hinge of matter converting into spirit is Monakos. . . . Monakos, at least *the* Monakos, goes about the business of reality with a firm grip, at times a self-destructive vise, on body-and-soul. It is a quickening of the body in direct touch with the raw physical-physiological make-do; a quickening of the soul that is none other than the quickened body transcending the physiological splendor it embodies, the uninterrupted miracle of the self-transcending food chain. In this context of a spirituality locked in the sweat of the flesh, the technology of evolution, I am going to elaborate on the environmental-architectural (arc-ology) work leading to Arcosanti and beyond" (133–34). The practice of building Arcosanti, therefore, becomes for Soleri a "neomonastic proposition" to counterbalance

material consumerism with mental transcendence. "To keep things in balance," he concludes, one can find in monastic models some important values, such as "tradition, learning, intellectual rigor, humility, frugality, altruism, self-discipline and transcendence," which are things "not to dismiss nonchalantly" (*Arcosanti: An Urban Laboratory?* 77).

36. For more discussion of these flows, see Robert Reich, *The Work of Nations: Preparing Ourselves for 21st-Century Capitalism* (New York: Knopf, 1991); and David Harvey, *The Condition of Postmodernity* (Oxford: Blackwell, 1989).

37. See Soleri, *Arcology: Architecture in the Image of Man*, 33–122.

38. Soleri seems to have faced these realities inasmuch as he has modified the presentation of Arcosanti as a planned community. Even the promotional imagery used to tout Arcosanti's future now has changed, perhaps in response to local real-estate industry expectations. In the late 1960s, Soleri's *Arcology: The City in the Image of Man* depicts a double assembly of two squat, solid masses, each resting on four major apses and several cylindrical columns, all providing cellular agglomerations of various working and playing spaces. Concrete, stone, and mortar appear to dominate the original design, as if Arcosanti was to simulate some timeless Italian hill town. As it has been built, what were to have been more integrated and compact spaces also have been disaggregated and opened, leaving what the first basic design saw as peripheral structures to become the main bulk of what now stands. Most recently, video programming and new three-dimensional models for "Arcosanti 2000" show an extremely glitzy, almost conventional complex of structures complete with smoked glass-sided high-rises and tubular external buttresses reminiscent of all the failed real-estate high-rise developments of the 1980s savings-and-loan debacles. It has the Soleri style, but this Arcosanti 2000 also could be something being puffed up by a Del Webb or a Charles Keating as *the* Arizona real-estate project of the next century.

39. For echoes of this vision of aesthetic dictatorship, see the writings of Soleri's American teacher, Frank Lloyd Wright, including his *An Organic Architecture: The Architecture of Democracy* (London: Lund, Humphries, 1939); *When Democracy Builds* (Chicago: University of Chicago Press, 1945); and *The Natural House* (New York: Horizon Press, 1954).

40. For more discussion of the "Arizona Viper Militia," see Steven A. Holmes, "U.S. Charges 12 in Arizona Plot to Blow Up Government Offices," *New York Times* (July 2, 1996), A1, 12; and Pierre Thomas and Serge F. Kovaleski, "Arizona Militia Group Depicted by Experts as Small and Secret," *Washington Post* (July 3, 1996), A3. Working dead-end, low-pay jobs as air-conditioning repairmen, used furniture salesmen, military surplus store clerks, doughnut bakers, strip-joint bouncers, and maintenance men, the Viper identity and com-

munity were tied entirely to owning, shooting, and accumulating firearms. With little else to occupy their time and energies in Phoenix as a city with little "urban effect," as one acquaintance observed, the Vipers "go out and blow up rocks in the desert and shoot their guns. They have been doing this for years. This is no secret" (cited in James Brooke, "Agents Seize Arsenal of Rifles and Bomb-Making Material in Arizona Militia Inquiry," *New York Times* [July 3, 1996], A18). See also James Brooke, "Volatile Mix in Viper Militia: Hatred Plus a Love for Guns," *New York Times* (July 5, 1996), A1, 16.

41. See Tim Luke, "Searching for Alternatives: Postmodern Populism and Ecology," *Telos* 103 (spring 1995): 87–110.

9. Community and Ecology: Bookchin on the Politics of Ecocommunities and Ecotechnology

1. Murray Bookchin, *The Ecology of Freedom* (Palo Alto, Calif.: Cheshire Books, 1982), 20–21.

2. Michael Walzer, "The Communitarian Critique of Liberalism," *Political Theory* 18, no. 1 (February 1990): 7. See also Alasdair MacIntyre, *After Virtue* (Notre Dame, Ind.: University of Notre Dame Press, 1981).

3. See Christopher Lasch, *The True and Only Heaven: Progress and Its Critics* (New York: Norton, 1991), 476–508. See also Alvin W. Gouldner, *The Future of Intellectuals and the Rise of the New Class* (New York: Seabury, 1979), 11–47; and Robert B. Reich, *The Work of Nations: Preparing Ourselves for 21st-Century Capitalism* (New York: Knopf, 1991), 171–84, 225–40.

4. Walzer, "The Communitarian Critique of Liberalism," 9. See also Robert Bellah et al., *Habits of the Heart* (Berkeley: University of California Press, 1985).

5. Lasch, *The True and Only Heaven*, 509–32. See also Christopher Lasch, "The Revolt of the Elites," *Harper's* 289 (November 1994): 39–49. For another, more critical, reading of the new class concept, see Barbara Ehrenreich, *Fear of Falling: The Inner Life of the Middle Class* (New York: Harper, 1989).

6. Murray Bookchin, *The Rise of Urbanization and the Decline of Citizenship* (San Francisco: Sierra Club Books, 1987), 5. Bookchin connects these forces to the dynamics of present-day urbanization that limit and check the truly civilizing benefits of creating and living in cities. Contemporary urbanization does not generate opportunities for building humanly satisfying cities or a humane civilization; instead, urbanization generates empty centers for new class managerial and financial transactions defined in terms of "a largely privatized interaction between anonymous buyers and sellers who are more involved in exchanging their wares than informing socially and ethically meaningful

associations" (Murray Bookchin, *Urbanization without Cities: The Rise and Decline of Citizenship* [Montreal: Black Rose Books, 1992], 8).

7. Ibid., 12.

8. Ferdinard Toennies, *Gemeinschaft und Gesellschaft* (Leipzig: Reisland, 1887).

9. See Henry Maine, *Ancient Law* (London: Murray, 1861); Émile Durkheim, *The Division of Labor in Society* (New York: Macmillan, 1933); Max Weber, *The Theory of Social and Economic Organization*, ed. Talcott Parsons (New York: Free Press, 1947); Charles Horton Cooley, *Social Organization* (New York: Scribner's, 1909); Ralph Linton, *The Study of Man* (New York: Appleton-Century, 1936); and Talcott Parsons, *The Social System* (Glencoe, Ill: Free Press, 1952).

10. Bookchin, *The Rise of Urbanization*, 4.

11. Ibid., 255.

12. See Murray Bookchin, *Remaking Society: Pathways to a Green Future* (Boston: South End Press, 1990), 19–39.

13. Murray Bookchin, *Toward an Ecological Society* (Montreal: Black Rose Books, 1980), 15.

14. Karl Hess, *Community Technology* (New York: Harper and Row, 1979), 14.

15. Bookchin, *Toward an Ecological Society*, 21.

16. Ibid., 37.

17. Bookchin, *Toward an Ecological Society*, 13.

18. Ibid., 47.

19. Ibid., 47–48.

20. Ibid., 110.

21. Ibid., 69.

22. Ibid.

23. Ibid., 69–70. This utopianizing quality has prompted criticism. As Alan Rudy and Andrew Light observe, "There is little in Bookchin's work that really attempts to theorize the problems of technological manipulation and control under capitalism as a unique form of determinism" ("Social Ecology and Social Labor: Consideration and Critique of Murray Bookchin," *Capitalism Nature Socialism* 6, no. 2 [June 1995], 75–106).

24. E. F. Schumacher, *Small Is Beautiful: Economics as if People Mattered* (New York: Harper and Row, 1973), 153.

25. Murray Bookchin, *Post-Scarcity Anarchism* (Berkeley, Calif.: Ramparts Press, 1971), 87.

26. Ibid., 112.

27. Bookchin, *Toward an Ecological Society*, 109.

Conclusion. New Departures for Ecological Resistance

1. For more discussion of the Nature/Denature dynamic, see Timothy W. Luke, "Liberal Society and Cyborg Subjectivity: The Politics of Environments, Bodies, and Nature," *Alternatives* 21 (1996): 1–30; and Timothy W. Luke, "At the End of Nature: Cyborgs, Humachines and Environments in Postmodernity," *Environment and Planning A* 29 (forthcoming 1997).

2. Bruno Latour, *We Never Have Been Modern* (London: Harvester Wheatleaf, 1993), 15. See also Ulrich Beck, *Ecological Politics in an Age of Risk* (Cambridge: Polity Press, 1995).

3. One response for ecocriticism is to resist the differentiation of Nature by ecocritics into analytically pluralized variants, such as first, second, and third nature, which prioritize one set of attributes in Nature over all others. Third nature's cybernetic telemetricalities, second nature's industrial territorialities, and first nature's biophysical terrestrialities all are often used to privilege some vision of Nature variously over others by different ecocritics. Third, second, and first nature are all natural, or autochthonous, phenomena that need to be integrated again as fully fundamental or obviously original equivalent expressions. All life on Earth is "natural," whether it is biomorphic, machinomorphic, or cybermorphic—a fact typically ignored by deep green ecology as well as shallow resource managerialist environmentalism. For more discussion, see Timothy W. Luke, "Placing Powers/Siting Spaces: The Politics of Global and Local in the New World Order," *Environment and Planning D: Society and Space* 12 (1994): 613–28; and Timothy W. Luke, "From Commodity Aesthetics to Ecology Aesthetics: Arts and the Environmental Crisis," *Art Journal* 51, no. 2 (summer 1992): 72–76.

4. See Jean Baudrillard, *Simulations* (New York: Semiotext[e], 1983), 23–26. In other words, Biosphere 2 is there for the sake of argument here to conceal the fact that *it* is now the reality of Biosphere 1. Its denatured terraformation is presented as imaginary to make us believe that Nature is real, when in fact the Arizona desert and all the rest of Nature surrounding it is now what is perhaps hyperreal.

5. See Timothy W. Luke, "On Environmentality: Geo-Power and Eco-Knowledge in the Discourses of Contemporary Environmentalism," *Cultural Critique* 31 (fall 1995): 57–81.

6. Ibid.

7. Ibid.

8. For a parallel argument, see Andrew Ross, *The Chicago Gangster Theory of Life: Nature's Debt to Society* (London: Verso, 1994); and Timothy Fridtjof

Flannery, *The Future Eaters: An Ecological History of the Australasian Lands and People* (Chatswood, New South Wales: Reed Books, 1994).

9. See Tim Luke, "Informationalism and Ecology," *Telos* (summer 1983): 59–73; David Dickson, *Alternative Technology* (Glasgow: Fontana, 1974); and Ivan Illich, *Energy and Equity* (New York: Harper and Row, 1974).

10. For more discussion, see William Leiss, *The Domination of Nature* (Boston: Beacon Press, 1974), and Enrique Leff, *Green Production: Toward an Environmental Rationality* (New York: Guilford Press, 1995).

11. See Murray Bookchin, *Toward an Ecological Society* (Montreal: Black Rose Books, 1980); and Timothy W. Luke, "Notes for a Deconstructionist Ecology," *New Political Science* 11 (spring 1983): 21–32. See also Steve Chase, ed., *Defending the Earth: A Dialogue between Murray Bookchin and Dave Foreman* (Boston: South End Press, 1991), for Bookchin's analysis of deep ecology and biocentrism as political programs.

12. See, for example, Martin Ryle, *Ecology and Socialism* (London: Radius, 1988); and Stuart Hall and Martin Jacques, eds., *New Times: The Changing Face of Politics in the 1990s* (London: Lawrence and Wishart, 1989).

13. Bookchin, *The Rise of Urbanization and the Decline of Citizenship* (San Francisco: Sierra Club Books, 1987), 13–14.

14. As Wendell Berry notes, some cities will never be sustainable: "New York City Cannot Be Made Sustainable, Nor Can Phoenix," but many others can be and should be. See "Out of Your Car, Off Your Horse," *Atlantic Monthly* 267, no. 2 (February 1991): 60–63. See also David Morris, *Self-Reliant Cities: Energy and the Transformation of Urban America* (San Francisco: Sierra Club Books, 1982); and Richard Register, *Ecocity Berkeley: Building Cities for a Healthy Future* (Berkeley: North Atlantic Books, 1987). More discussion of these difficulties can be found in Martin O'Connor, ed., *Is Capitalism Sustainable? Political Economy and the Politics of Ecology* (New York: Guilford Press, 1994).

15. See an initial consideration of these goals in Ken Anderson et al., "Roundtable on Communitarianism," *Telos* 76 (summer 1988): 2–32. These ecological populist communities would work best in institutional arrangements for promoting face-to-face participatory democracy or localistic confederal institutions. Confederalism, in Bookchin's analysis, "is above all a network of administrative councils whose members of delegates are elected from popular face-to-face democratic assemblies, in the various villages, towns, and even neighborhoods of large cities" (*Urbanization without Cities: The Rise and Decline of Citizenship* [Montreal: Black Rose Books, 1992], 297). A related discussion of the democratic challenge of such systems is Daniel Press, *Democratic*

Dilemmas in the Age of Ecology: Trees and Toxics in the American West (Durham, N.C.: Duke University Press, 1994).

16. For more discussion, see Paul and Percival Goodman, *Communitas: Means of Livelihood and Ways of Life* (New York: Random House, 1960). See also Wendell Berry, *Meeting the Expectations of the Land: Essays in Sustainable Agriculture and Stewardship* (San Francisco: North Point Press, 1984); and Wendell Berry, *The Unsettling of America: Culture and Agriculture* (San Francisco: Sierra Club Books, 1977).

17. See Murray Bookchin, *Remaking Society: Pathways to a Green Future* (Boston: South End Press, 1990), 127–204.

18. See Bookchin, *The Rise of Urbanization*, 123–73.

19. E. F. Schumacher, *Small Is Beautiful: Economics as if People Mattered* (New York: Harper and Row, 1973), 154; emphasis in the original.

20. See Ivan Illich, *Tools for Conviviality* (New York: Harper and Row, 1973).

21. Amory B. Lovins, *Soft Energy Paths: Toward a Durable Peace* (Cambridge, Mass.: Ballinger, 1977), 57.

22. Resisting new class administrative domination would require confederalist countermovements to rebuild material arcologies around ecotechnologies and rethink symbolic systems as ecocommunities centered on the active rational citizens in each local community. As Bookchin argues, confederalist communities are the essential ensemble of key values: "Decentralization, localism, self-sufficiency, interdependence—and more. This more is the indispensable moral education and character building—what the Greeks call *paideia*—that makes for rational active citizenship in a participatory democracy, unlike the passive constituents and consumers that we have today. In the end, there is no substitute for a conscious reconstruction of our relationship to each other and to the natural world" (*Urbanization without Cities*, 299).

23. For some new class articulations of other possible futures for transnational corporate capitalism and traditional nation-states, which would resist Bookchin's ecological communitarianism, see Robert B. Reich, *The Work of Nations: Preparing Ourselves for 21st-Century Capitalism* (New York: Knopf, 1991); Lester Thurow, *Head to Head: The Coming Economic Battle among Japan, Europe, and America* (New York: William Morrow, 1992); and Paul Kennedy, *Preparing for the Twenty-First Century* (New York: Random House, 1993). Christopher Lasch expertly illustrates the oligarchical anticommunitarianism and postnationalism of the symbolic analysts thriving amid these new class formations in "The Revolt of the Elites," *Harper's* 289 (November 1994): 42–46.

Index

Timothy W. Luke is a professor of political science at Virginia Polytechnic Institute and State University. He also has taught at Victoria University of Wellington (New Zealand), the University of Missouri-Columbia, and the University of Arizona. He is the author of *Shows of Force: Power, Politics and Ideology in Art Exhibitions* (Duke University Press, 1992), *Social Theory and Modernity: Critique, Dissent, and Revolution* (Sage, 1990), *Screens of Power: Ideology, Domination and Resistance in Informational Society* (University of Illinois Press, 1989), and *Ideology and Soviet Industrialization* (Greenwood, 1985).